LARGE-SCALE CONSTRUCTION PROJECTS

LARGE-SCALE CONSTRUCTION PROJECTS

MANAGEMENT, DESIGN AND EXECUTION

Salman T. Al-Sedairy

Batsford Academic and Educational

London

Typeset by Keyspools Ltd, Bridge Street, Golborne, Lancs
and printed in Great Britain by Butler & Tanner Ltd
Frome, Somerset

Published by Batsford Academic and Educational
an Imprint of B. T. Batsford Ltd
4 Fitzhardinge Street, London W1H 0AH

British Library Cataloguing in Publication Data

Al-Sedairy, Salman T.
   Large scale construction projects: management,
   design and execution.—(Mitchell's professional
   library)
   1. Construction industry—Developing countries
   —Management   2. Industrial project management
   —Developing countries
   I. Title   II. Series
   624'.068'4   HD9715.D442

ISBN 0 7134 51289

CONTENTS

Scope of work
   Task identification
   Task management analysis
   Selection, assignment and contracting
   Implementation and monitoring

LIST OF FIGURES

FOREWORD

by Professor David Gosling

Development in the Third World divides itself sharply into two categories. On the one hand, there are enormous problems of poverty, as well as social and economic change, which countries in Africa, South-east Asia, the Indian subcontinent and Latin America face; but on a different level, there are the problems of rapid expansion in wealthy states developing as a result of an oil-based economy. There has been a tendency in both categories, fostered by international development and relief agencies, to utilise the expertise of North American and European technical consultants in dealing with these difficulties. All too often, this expertise is misplaced in not adequately recognising the historical, cultural, social and climatic framework of the countries seeking assistance. Much has been written by western sociologists and others concerning the negative impact of rapid and misdirected development in the Third World. The role of these writers has, however, been largely founded in negative criticism rather than the assessment of positive alternative approaches.

In the Middle East, surprisingly little has been written about development methodologies. There are complex religious, cultural and climatic issues which need to be understood. After the Energy Crisis in 1973, with the growing importance of oil as a fossil fuel and basic raw material in many manufacturing industries, many of the Middle Eastern states undertook ambitious programmes of social change and industrial development. Much of the early work was undertaken by multinational companies, but in recent years expertise in building and design has grown considerably in the Arab states themselves. To meet the continuing challenges of development and of the improvement of the living environment, it is essential to understand the complexities of managing building programmes on the largest scale.

The author, Dr Salman Al-Sedairy, is a young Saudi Arabian architect with wide experience of the nature of the problem. Formerly Chief Archi-

tect and subsequently Director General of Construction at the Ministry of the Interior in Saudi Arabia, Dr Al-Sedairy received his professional education in Saudi Arabia, the United States of America and the United Kingdom and shows his unique and extensive knowledge of the design and building process in Saudi Arabia.

This study, based upon a doctoral dissertation, differs from many American and British publications on construction management. It presents the problems and solutions in construction management from the point of view of the client, possibly for the first time. In an erudite and precise style, the author analyses the system of policy-making concerned with all aspects of planning, design, construction, operation and financing of large scale building projects. One of the interesting aspects of this particular study is that, in putting forward a systematic approach to planning, estimating and design, it is recognised that there are a number of equally valid approaches, depending upon the particular circumstances obtaining at a particular time.

The book could reasonably be considered essential reading for American and European architects and engineers who have been or are currently involved in construction in the Middle East. It is of equal importance to major contractors, subcontractors and their site-management personnel. A unique insight is given into the complex process of consultant and contractor competitive selection. It is also a key reference work for the client organisations themselves, providing a clear guide for personnel involved in policy-making and decision-making who are aiming to ensure much higher constructional standards and safeguards in the future.

This is not to say that the study is merely of relevance to the powerful and affluent oil-producing countries and their governments. It should be required reading for personnel in international relief and technical assistance agencies responsible for Third World Development. A rational approach to project management methods and monitoring techniques might do much to avoid the failures and disappointments of many of the well-intentioned but sometimes misdirected development initiatives in the poorer countries.

Dr Al-Sedairy's interest in construction is not confined to management techniques alone. As author of two previous books, the first on urban design and community development in Saudi Arabia and the second on landscape architecture in Saudi Arabia, he adopts a fresh approach in emphasising that the goals are not merely the achievement of building programmes and target costs but also of high standards of architectural design and environmental improvement.

In detail, the study itself outlines European and American management techniques, including Base Line Studies as well as Project Control. Methods of monitoring performance, project control systems such as the assembly of essential information, and the communication process in project management are described. The book takes the reader through consultant selection processes, schematic design preparation, design development, and design management systems. Construction tender procedures now in use, with special emphasis on tender evaluation and contract award, are listed. Different types of competitive bid contracts and negotiated contracts are analysed as well as general construction supervision, together with maintenance and operation of completed projects.

All available information has been assembled in a comprehensive way, which, together with extremely clear diagrams providing concise summaries of the systems described, transforms the study into a seminal work on project management and design. All present evidence suggests that such a book is needed in the great majority of developing countries and has direct relevance to the problems of rapidly expanding Third World cultures. The main challenge was to order and clarify the complex issues involved and to formulate effective courses of action. This has been achieved in an exemplary way.

There are many recently published books dealing with design in the hot, arid lands, but the majority analyse design responses to climate and, occasionally, to the cultural and religious aspects of architecture. The present work is quite different, in that it is a manual of information and instruction which has direct and immediate application to the problem of the management, design and execution of large scale constructional projects. Such a manual has been needed for many years.

David Gosling
BA(Arch), DipTP, MArch(MIT), MCP (Yale), RIBA, ARIAS, MRTPI, FRSA

Department of Architecture
University of Sheffield
United Kingdom

March 1985

ACKNOWLEDGEMENTS

The author wishes to express his sincere appreciation to Professor David Gosling for his advice and assistance, also for the assistance of Wendi Jarvis and Andrew Aucock.

1
INTRODUCTION

The development of a large scale building programme consists of four basic segments:

1. Project inception
2. Design
3. Construction
4. Operation and maintenamce.

Each segment requires a different operations approach by different members of the project development team. The members of the team will have a variety of expertise. For example the executive manager's involvement will be reduced during each phase of a project until the user of the project needs little or no involvement by the project development team.

A large scale building project can also be thought of as having four basic components:

1. Participants
2. Information
3. Processes
4. Products.

The government executive and his project development team become the catalyst which ensures that participants, process and information gathering achieves the correct objectives.

The distinction between the four basic segments and the four basic components is that the segments normally work best in sequence and each sequence is distinctly different from the other. The components are always overlapping, running in parallel, and intertwined together. The government's task is to see that these overlapping, parallel and intertwined components work in harmony in each phase of the four basic segments. The tool which creates this harmony is good management and appropriate policies.

The base line study guide for government executives

The foundations for the policies and methods of managing a large scale project are described in the two parts of the project inception segment. These two parts are Base Line Study Part I and Base Line Study Part II. Part I determines the nature of the project and Part II determines the method of production. With this base line study guide and a correct decision-making process, a successful project can be produced by the government executive and his team without the need for an outside consultant. It is more likely that an aware government agency will produce the most satisfactory project in terms of the future user, who is likely to be represented by some other government agency.

The difference between a successful and unsuccessful project relates to decisions and to resources. If comprehensive project information is put forward in a clear manner and if the alternatives are correctly analysed, the 'concerned' decision-maker will make the appropriate decisions. If the decisions are made by the less concerned or those with priorities different from a successful project, the decisions could lead to project failure.

For the 'concerned' decision-maker who desires to make the appropriate decisions for a successful project, the base line study is dependent upon a 'systems approach'.

The way to optimise the skills, and execute a large scale project, is through experience and the accumulation of expertise. Such experience is not gained by merely working on one aspect of a particular project but rather by obtaining a comprehensive overview of the project as a whole. In the first instance the initial appraisal of information makes the task appear more formidable than it really is. For example, it is likely that after reading and absorbing instructions for the operation of a video recorder or calculator in everyday life, the operation of the equipment itself may appear comparatively easy compared to the initial understanding of the operating instructions. Using a simple base line study guide takes the operator through the basic steps in the operation of any successful project.

Major pioneering government programmes which received international acclaim and involved the systems approach were the California School Construction Systems Development (SCSD), the New York Construction Fund (SUCF) and the Illinois Capital Development Board (CDB) in the USA.

The SCSD and SUCF were involved in more specialised project types while the CDB dealt with multi-type building projects. Nowadays most United States government organisations follow these models. The organisation is similar in scale to a Saudi Arabian ministry building development organisation.

One of the most successful examples of the application of these techniques appears in the execution of large scale building projects in Saudi Arabia. Many of the initial large scale projects were developed in the last two decades and were not as successful as the more recent projects of the last four years when Saudi government building agencies' executives started to lead their own project development teams. Multi-million dollar project failures can be seen all around the world in the form of obsolete airports and hotels, commercial centres with inadequate parking, new towns behind programme, and mass high-rise housing inappropriate to the needs of the users. Mass high rise housing is a clear example of a project which could have been successful through the use of Base Line Study I which is intended to investigate the true purpose of a project. Most recent airports, new community developments, and government facilities are generally built on programmes within budgets, and especially with the object of serving the needs of the user better than preceding projects. Like the examples in the United States and the United Kingdom, concerned Saudi government project teams are now using a systems approach and base line type studies to produce more successful building projects than those hitherto managed almost exclusively by foreign (international) project teams.

Project inception

Every project must have a starting point, whether it is merely an idea in the mind of a potential user who needs a facility or whether it is a government agency acting under instructions to plan for a facility to meet a particular future need. The greatest concern of the government executive, who has the task of commencing operations for a project, is establishing the starting point and the final objectives. The project inception section provides an outline guide, policy and insight to answer these questions. This section includes:
- Base Line Study I – Statement of Client Needs
- Base Line Study II – Project Implementation Plan

The project inception section describes all segments and components needed before the design of a facility can start.

1. It involves the participants in information-gathering. The participants would include client-user, project developer, project programmer, site acquisition team, and the design consultant upon appointment.
2. It includes the information necessary for the preparation of the project: client goals, project information, project management requirements, site data and design consultant requirements
3. It provides the information necessary for both the product and the participants: problem statement, project implementation plan, funding proposal, consultant prequalification forms and invitations for proposals, and site acquisition requirement problem statements
4. It produces the essential items required before design can commence: funding and site.

The detailed procedures are not included in this book but can be developed for a specific project from the policy statements and outline guides. The Base Line Study II describes the type of management plan for design, construction, and the operations and maintenance phases. Within each phase the management outline and policy are given. The management plan, outline, policy and detail procedure would be included in the actual project in the Base Line Study II, referred to as the Project Implementation Plan.

Design

Design is an abstract notion which raises many questions. Is design magic? Is it creativity? Is it instinct? Is it something that only an architect or engineer can visualise? For many centuries architects, engineers, and planners have been involved in the creation of many structures, urban complexes, and communities throughout the world. They thought about urban prospective enclosures made from local materials, the size of spaces formed by these enclosures, and the spatial relationships. This could be considered to be the magic: creativity, instinct and visualisation.

Today there are specialists who design cars, televisions, and other complex technology. If the design of one of these items satisfies the criteria of appearance, function, durability and price it will be commercially successful. In building terms the

2

client-user knows the spaces needed and the spatial relationships required (programme); as well as possessing a response to appearance (schematic); the approximate budget (cost estimate); and when it is required (time schedule). The architect and engineer must make sure that the requirements are explicit and justified (design development). These requirements must be communicated in a clearly understood language to the contractor who erects the building (construction documents).

From this process comes the traditional linear method of design. But where constraints of time and programme become important other methods have evolved. The fast track method is a process of overlapping designs with construction to shorten the delivery time of a project. The turnkey method was developed to allow one agency to think for the client user and to manage and control the project as well. This method was again derived from the constraints of cost and time. Constraints can vary from the project delivery process and thus affect the method of design. The traditional linear method of design is presented as one of the best possible methods of satisfying the requirements of the project participants: client-user; designer; contractor; as well as the operator of the finished project.

Variations on the linear method are clear. The phases of design are outlined below with an indication of their importance and benefits. These phases include:

1. Programme development (understanding client-user needs)
2. Schematic design (visual image of client-user needs)
3. Design development (definitive design)
4. Construction documents (translation of client-user needs for contractors)

Construction

From the design comes the ideas, from the construction comes the reality. Are project needs properly defined? (Base Line Study I): Are the project needs properly documented ideas? (design): Is the construction examined to assure that the documented ideas are built (construction supervision)? If these criteria are met the project is likely to be successful. If one element is ignored, the construction (involving the largest expense, longest time, and greatest effort) is wasted.

In the construction phase, the most important factors are:

1. Appointment of a qualified builder within the budget limits prequalification
2. Constant inspection of design during construction to ensure compliance with contract documents
3. Ensure the contractor's cash flow (payment processing)
4. Ensure that the project is satisfactorily completed and handed over to the client-user in such a way that operations and maintenance can be a satisfactory process (handing over).

Operation and maintenance

If a facility cannot be operated and maintained continuously, there is an obvious waste of money and effort to the detriment of the user. Operation and maintenance is seen as a priority and the justification of the Base Line Study I.

If the Base Line Study I is satisfactory operation of the building by the user can be carried out efficiently without instructions or supervision.

Typical of lack of care in the initial design studies are buildings in Jeddah with basements continually under water, or unoccupied high-rise housing in Riyadh, or facilities in Dhahran which cannot be maintained adequately by local inhabitants.

There must therefore be an outline of operation and maintenance which is presented under the following headings:

1. Tendering for services
2. Preparation for operational services
3. Technical operations
4. Technical maintenance
5. Non-technical operations and social services

Analysis and conclusions

It is obvious that knowledge improves with experience. The feedback and analysis system is emphasised so that the project developed can remain on programme and difficulties may be avoided in future projects. A written analysis of any segment of the building industry should include, for each major project, the 100 best and 100 worst elements of the project and project process.

2
PROJECT PROCESS AND MANAGEMENT

The principles of management

In the following chapters, it is the intention of the author to describe in detail a viable management process for large scale projects and to relate this to the particular experience in the Kingdom of Saudi Arabia. It is important first of all to establish the meaning of the term 'management'. Management may be defined as 'the creation of conditions to bring about the achievement of objectives by the optimum use of available resources' (RIBA 1980), though it must be realised that it is people and not the process that brings about this achievement. As Peter F. Drucker points out, 'People manage rather than "forces" or "facts". The vision, dedication, skills and integrity of managers determine whether there is management or mismanagement.' (Drucker 1974) Management is a constant state of dynamics which should be independent of rank, power or ownership, where people are brought together to accomplish a goal. Management is a discipline in which every manager has the common goal of bringing effectiveness and efficiency to a task, project or organisation. Effective management can be achieved in many ways, depending on the objectives of the tasks at hand. This research will review the methodology and tools of project management inasmuch as they relate to the management process of large projects.

Project management: the systems approach

With the advent of the industrial revolution and the widespread application of scientific thought, building types and user requirements increased in complexity. As a result, it became apparent that the conventional methods of constructing the built environment were inadequate. Albert J. Kelly cites these reasons for the increased complexity of projects in recent times:

While the increase in dollar value clearly owes much to nearly a decade of high inflation (a $400 million project of ten years ago would now cost over $1 billion on the basis of 10% inflation per year), there is undoubtedly an underlying complexity of factors pushing towards larger and more expensive projects. Today's population is large and growing, in consequence so is the demand for material growth; technological sophistication is no longer a luxury but often a necessity, and it is expensive; material growth and technological sophistication brings technical, economic and regulatory complexities, and these add further costs; the social costs of not investing in large development projects are very high, governments often recognise massive development projects as unavoidable. (Kelly and Morris 1981).

As users, owners and government agencies' requirements have grown more complex, there has been more demand placed on the design profession to be responsible for the time, scope, cost and quality of a project. Later, when better education heightened the critical awareness of the end-users, and the environmental crisis and the realisation that there was a limited supply of natural resources in turn developed a concern for a more efficient and cleaner use of energy, designers were forced to deal with a greater and more complex range of issues. There were growing concerns about the environmental impact a particular project might have, not only on a local scale but at a national level. More and more research was developed which brought up an ever-increasing range of issues which had to be resolved.

During the mid-sixties projects grew in such complexity that conventional methods of completion grew chronically inadequate. In response to this inadequacy, two programs were developed in the United States, The California School Construction Systems Development (SCSD) and the New York State 'Construction Fund (SUCF). Both programs were initiated to develop a mechanism to manage effectively the complex issues of time,

4

quality and cost of a project. Both systems were essentially the same and provided a vehicle for project managers to take a systematic or rational approach to problem solving, programme management and project development (Jacques 1976); hence, the name, 'systems approach'. The SUCF and SCSD systems evolved into a management-intensive approach aimed at solving problems in a systematic way and thus allowing more effective usage of time, materials and costs; also, through a series of evaluation methods, to develop projects which effectively dealt with the original goals and objectives of the owner. The 'systems approach' allowed the designers to:

1. increase concern for and commitment to the user needs as a basis for design
2. develop reliable methods for identifying and choosing alternatives in planning, design and construction
3. use sophisticated performance criteria and performance specifications as a means of defining programme goals and objectives
4. utilise comprehensive long-range programme planning and physical planning procedures
5. apply comprehensive construction management skills to assist in meeting limited budgets and schedules
6. use feedback and evaluation techniques as a basis for improving building programme performance (Drucker 1974).

Over the years, there has been a multitude of variations to the SCSD and SUCF programmes, each developed for a particular project application.

In the early seventies the Illinois Capital Development Board (CDB), after two years in operation, were using the 'systems approach' to develop higher quality projects, at less cost and in less time, with less paperwork and less staff, than any major governmental or private consultants were doing with similar scale projects.

Any project management system, in order to work effectively, must be able to achieve the following:

1. There must be a method of programme analysis and definition.
2. There must be a system of evaluation in order to determine alternatives and options against a set of predefined goals, objectives and needs.
3. There must be a process of project delivery.
4. There must be a system of evaluation and feedback to provide data for improving performance in the future.
5. Each method, procedure or tool developed

must ensure control over time, cost and quality of a project. This last item is a criterion for every stage of the project management system.
6. The system must be flexible enough to allow for change and be adaptable to cater for present unknown and unforeseen circumstances (Morris 1973; Kelly and Morris 1981).

It must be realised that the 'systems approach' is a methodolgy developed in conjunction with appropriate equipment to deal with complex project planning and delivery in a logical way. It must also be realised that management '... is the essential, creative tool which is utilised to assist in the attainment of complex programme and project goals.' (Jacques 1976).

The Project Manager

Project management has two main aspects. On the one hand there is the process, and on the other hand are personnel. In large scale projects, hundreds or even thousands of people may be involved in the project management process and perhaps tens of thousands involved in project delivery. At the heart of any project management process is the project manager.

There are other titles for this type of person such as Project Director, Project Co-ordinator, Project Head, Principal-in-Charge; but the best way he can be identified is by the role he plays in the project management process. For clarity, an analogy can be drawn with the conductor of a philharmonic orchestra, where the the conductor is the co-ordinator, and the different instrument sections may be likened to project design teams, owner's representatives, specialist consultants and so on. In the symphony they are to play, each team has a designated and important part (the project delivery process). The Project Manager is an orchestrator, a central link between all the different groups involved in the project. It is with the Project Manager that the owner's needs and requirements are identified, planning and scheduling of the project are done, the systems of control of cost/time/quality are established and procedures and staffing requirements are developed.

In large scale projects, the role of the Project Manager can be duplicated many times at different stages and at different levels. In his proposal to the Ministry of Interior for the management of Phase I of the Internal Security Forces Housing Projects, for example, the author recommended the following management structure:

The top management team reports directly to the Ministry of Interior. It will be composed of two project directors, one of whom represents the staff of the Ministry, the other represents the Consultants' staff. Together they will make executive level decisions related to the project and monitor the project achievement on a periodic basis. Supporting their efforts will be deputy project managers from the staff of the Ministry and a team of four deputy project managers for the Consultants' staff. The responsibility of each of the four deputy project managers will be allocated to a regional area: North Sites, East Sites, South Sites and West Sites. (Al-Sedairy 1981).

The Internal Security Forces Housing Project is a very large scale project. Phase I entails the construction of approximately 30 communties on 30 different sites in the Kingdom with a total of some 12,000 housing units. The project was of such a scale that organisationally it was managed as a series of specialist building types (housing, community facilities, educational buildings, infrastructure etc.) each with its delivery schedule, related to the delivery schedule at a particular site. This was then co-ordinated with delivery of Phase I (of which there were three anticipated phases for the project) as a whole. In large scale projects it is often necessary to develop this type of management hierarchy for effectiveness and efficiency. Nevertheless the role or function of the Project Manager remains the same however circumstances may change.

In large scale projects there may be a dichotomy in the management process. There is the need to break down the management of tasks into digestible elements, and then interlink them in logical sequence. This can develop into a very complex process. On the other hand, management needs to be so structured as to allow a clear overall comprehension of the project as a whole. In his involvement with all the elements of the project delivery process, the Project Manager plays a particular role. There must be a clear distinction between those who carry out instructions and those who establish the instructions. It is the role of the Project Manager to develop the correct conditions for efficient execution and to assure that what needs doing is done according to plan. David Haviland describes these functions as the basic functions of management (Haviland 1981):

1. Planning – setting goals and objectives, establishing strategies and plans for accomplishing these objectives.
2. Organising – establishing a structure of activities, roles, responsibility and authority for carrying out objectives.
3. Staffing – identifying, assigning and training the people who make up the organisation.
4. Directing – providing day-to-day guidance and supervision of the staff.
5. Controlling – ensuring that the work being done conforms to plans, measuring performance and progress against plans, and correcting any deviations as they arise.
6. Evaluating – assessing overall performance of past and present efforts so as to improve future efforts.

The role of Project Manager in relation to government agencies

The role of the project manager in relation to a government agency is a function of that type of agency. It can be expressed with the formula:

$$PM = f(GA)$$

Where PM is the project management process, f is the function, and GA is the type of government agency. As the agency type changes so does the project management process. In most cases, the Project Manager will generally deal with one of two types of government agencies, technical or non-technical.

A technical agency can be described as an agency which, as part of its primary function, can knowledgeably develop its own programme requirements and is experienced at supervising the project delivery process. Examples of these would be the US Department of Housing and Urban Development (HUD) in America or the Department of the Environment in England, or the Ministry of Housing and Municipalities in the Kingdom of Saudi Arabia. Non-technical agencies are agencies whose primary functions are in other areas and who have no real experience in delivery of the built environment. Examples of these types of agencies in the United States would be the Environmental Protection Agency, Federal Aviation Agency, and Federal Drug Administration; and in the Kingdom of Saudi Arabia, The Ministry of Interior, the Ministry of Finance, the Ministry of Education.

When dealing with a technical government agency, the Project Manager's role is somewhat simplified. His client has a knowledge (or should have) about the project process and delivery. In most cases, the programming effort or base line study is already developed.

This does not mean that the starting point with this type of client is at a schematic design or 'synthesis' stage. On the contrary, careful analysis of the programme developed by the client should occur to ensure that recommendations are indeed realistic with regard to the services required of outside consultants, and whether the preliminary budgets and schedule set are realistic or feasible.

Most governmental departments tend to be organised along functional lines. The Government of the United States is a good example with its agencies, each performing a different function in the governing of the country. In Saudi Arabia it is much the same, with the tasks of governments divided among ministries, each performing a different function, e.g., Ministry of Defence and Aviation, Ministry of Agriculture and so forth. This division by function also occurs within the organisation of individual ministries.

Implementing government projects will involve functional departments. However, these departments are geared to performing and managing tasks that are continuous by nature. Their managers are used to working with routine tasks on a continuous basis with the only constraint being a yearly budget. A project, on the other hand, has a set of goals and objectives to meet within a specified period of time. Projects are usually unique, with the prediction of ultimate time and resources difficult.

In implementing a project that involves existing functional departments, the project manager has the responsibility of utilising the resources of these departments while satisfying the goals and objectives of the project. His objective is 'to produce a specific result within the technical, cost, quality and schedule specifications with the available resources of the organisation'.

The project process

Throughout history, the normal progress of a project was a linear process. Within this process, the architect would determine the client's needs, create a design, seek approval, develop all necessary construction documents, seek tenders and then build. Typically, the programme would be completed before the design began. The design was completed before the project was priced and prices were determined before construction began. (See Fig. 1 overleaf).

The traditional linear process has the advantage of control. As each step of the process occurs in sequence, there is the ability to review and approve each step before proceeding to the next. There is also control over cost, because the contractor bases his cost on a complete set of construction documents. The possible disadvantage is that, with no overlap, this process will take a longer time than other methods.

As owners now exert more pressure on the architecture/engineering profession to deliver projects in a shorter time, alternative methods lead to the 'fast-track' process. This process allows an overlap to occur in the design and construction phases of the project. Upon approval of the schematic design, the project is sub-divided into a number of tender groups which are developed and

then let in a pre-determined order according to construction logic. This shortens the delivery time of the project, allowing for early occupancy and a shorter time liability in terms of on-site construction financing or funding, thus saving excessive interest charges. Fast-track is sometimes called 'phased design and construction'.

The obvious advantage of the fast-track method is the saving in time. The system is the result of inflation in building and interest costs over the past decade, and the only feasible response to this is to reduce the time between the decision to build and the delivery date. There are disadvantages, in that construction will dictate the amount of time available for design. Before the design is finalised, critical decisions have to be made about long-lead items to allow design and implementation to occur as the construction sequence dictates. This means that decisions will be made in anticipation of their future effect. Such decisions rely upon previous experience of the system more than would be the case if the linear method were utilised. Phased sequence of design/construction, and the tender for separate packages, suggest that fast track requires a more sophisticated management structure. Construction management services are recommended for this method, together with systems of project control, including computer utilisation, making the management structure even more complex.

The project delivery methods

Construction involves three executive groups: the owner, the design team and a construction team. The construction team is often referred to as the General Contractor, and 'General Contracting' as the building method. The owner selects an architectural/engineering firm (A/E) to prepare construction documents and then engages a general contractor for the total execution of the project. The general contractor, acting as construction manager, sub-contracts various elements of the project but differs from a turnkey contractor in that the general contractor carries out some of the construction work with his own employees.

With the increasing pressure on programme times already mentioned, which affect such key items as quality and cost, the architectural engineering profession in turn has searched for alternative methods of control to replace the traditional linear process and general contracting. Some methods (e.g., 'fast-track' or 'design-build') respond to the primary determinants of a project by allowing an overlap in the delivery process, thus providing the flexibility to adapt to the different needs of an owner. (See Fig. 2 overleaf).

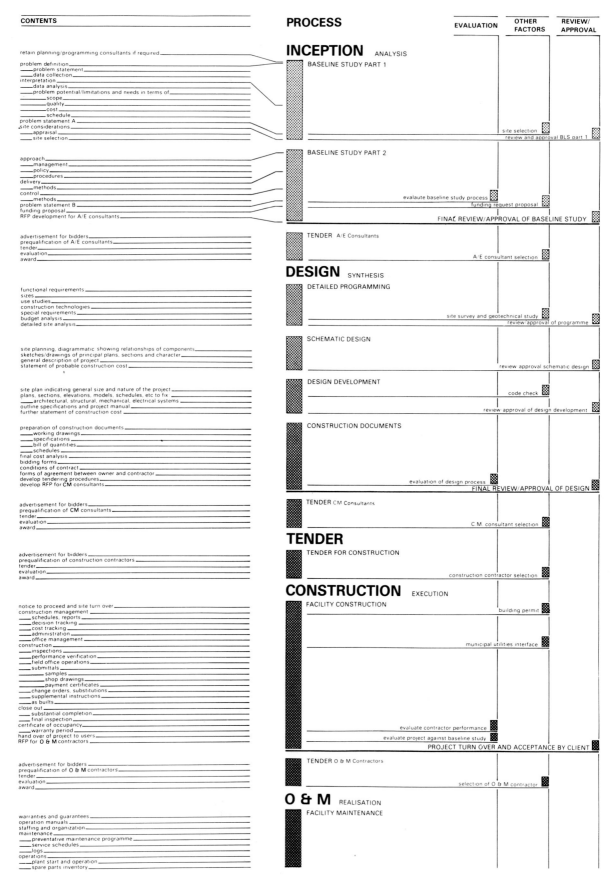

Fig. 1: PROJECT ACTIVITY DIAGRAM.

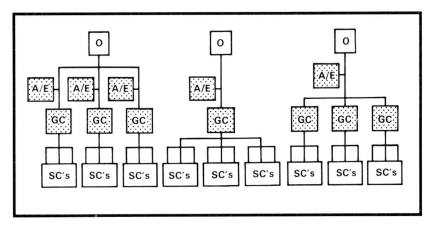

**GENERAL
CONTRACTING**

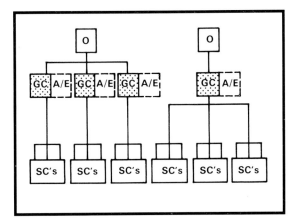

**DESIGN / BUILD
CONTRACTING**

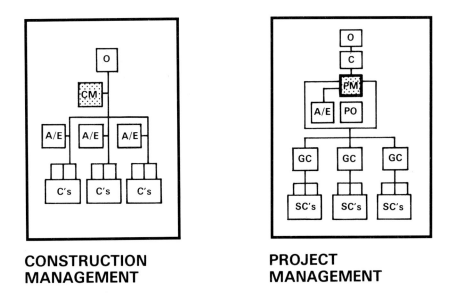

**CONSTRUCTION
MANAGEMENT**

**PROJECT
MANAGEMENT**

Fig.2: PROJECT DELIVERY METHODS.

Design-build delivery method

Design-build is essentially an integrated process. The programme is developed to a stage where it becomes sufficiently detailed for pricing to begin. Tenders are sought based on the programme for both the design and construction of the project. The contractor will be responsible for developing the design and the execution of the project and will (most of the time) sub-contract major parts of the work to design firms and construction sub-contractors. The single point responsibility is intended to allow better co-ordination between design and construction and obtain time and cost savings.

Design-build has the advantage of completing the project in a shorter period of time; however, there is the disadvantage that the owner loses a degree of control. In design-build a conflict of interest must always exist, whoever develops the design, for that person will constantly have to balance the interests of the owner against the profitability of the contractor. If, for instance, an A/E firm acts as the prime contractor sub-contracting the construction to a general contractor the A/E firm, when making any decisions, will have to resolve the conflict of the 'best choice for the owner' and 'the most profitable choice to the contractor'.

This method is sometimes confused with the turnkey method, but the turnkey contractor does none of the construction work with his own staff.

Construction management method

Professor Adrian, a construction management consultant, defines this as 'a process by which a potential project owner engages an agent, referred to as the Construction Manager (C.M.) who co-ordinates the entire project process, including project feasibility, design, planning, letting, construction, and project implementation, with the objective of minimizing the project time and cost, maintaining the project time and cost, and maintaining the project quality' (Adrian 1981). The author believes that this definition, as well as the method, is frequently used in developing countries to sell what could be termed 'contract management' methods, where the contract manager receives a commission for administering a project but without providing any design or construction services: it is an attractive concept but might not be in the interests of the owner. Such a method might be more feasible in industrialised countries because the national constructional system allows a greater degree of control than in a developing country. The C.M. staff retains powerful authority without any real responsiblility.

Project management method

The author has developed this particular method, which is a combination of the C.M. and general contracting, and has found it to be the most successful method in the administration of large scale projects.

The owner will develop an organisation which includes himself and the consultant personnel working together as one team, with the Project Director as the team leader. This team develops the design, based on the Base Line Study I, and then prepares construction documents for tender.

A different General Contractor will be selected to implement a different site or a different part of the work. The selection of the General Contractor should be restricted to those firms capable of executing a major percentage of the work with their own staff (normally not less than 60%).

Such a delivery method could be adopted to serve either the linear process, or fast-track process. The advantages and disadvantages of this method are discussed below.

1. The Project Director
Large scale projects are complex in scope, and in manpower resources as well as supervision.

Such projects need a complex technology with a large number of personnel from various disciplines and will require many interfaces. It follows then that project directors must have both technical and administrative expertise.

However, depending upon scope, complexity and peak manpower requirements, the technical or administrative functions may vary. For example, a complex project with numerous technical and management inputs and interfaces requires a strong administrative performance. This is the opposite of a single-discipline project with a small number of project personnel, where the project director may also function as chief engineer. The project director has to have a strong personality, the ability to plan, organise, lead, direct, control and make decisions, but also understand the single point of integrative responsibility.

2. Owner project personnel
Contributions must be defined and integratively planned and controlled during all phases of the project life cycle. Personnel should be carefully selected for honesty and commitment. If they are not, the project will be a disaster and participants will be able to hide behind this weakness.

3. Training programme
The success of the project may depend on an intensive training programme which should be developed through the life of the project.

The Project Organisation Plan

The owner must set up an organisation to develop the project at an early stage, with the establishment of a main executive committee. (See Fig. 3 overleaf). The Chairman of this committee should hold a senior position, in the owner's organisation.

Although there will be a main owner for a large scale project, various sections of the owner's organisation will have responsiblility for various aspects of the project. Each section of the owner's organisation would normally appoint a representative on the main executive committee, and each representative must have an alternate to act on his behalf. Other related departments, such as Financial Department, Budget Department, and Technical Department, would be represented on that committee.

The function of the committee is to establish overall policies and guidelines. The committee would select the Project Director and appoint him as a member of the committee. This Project Director would be responsible for representing the owner with all other parties, and vested with authority to enable him to manage the project in an efficient manner. Direct communication should be established between the Chairman and the Project Director. The Chairman provides interpretations of project policies, guidelines and advice to the Project Director as required.

The Project Director is responsible for the overall management of the project. Although the Committee is responsible for overall policy on any project, it delegates executive powers to the Project Director within clearly defined terms of reference. For instance, the Project Director is responsible for selecting the project team and the project team must have personnel capable of taking executive decisions in detailed planning.

Since it is extremely unlikely that the owner will employ the full range of professional disciplines necessary to develop the design and supervise the construction within his organisation, the selection of the A/E consultants is of prime importance. The Project Director will be guided by two considerations. The first is to expedite work efficiently and to a high standard. The second is to strengthen local expertise so that future projects can be undertaken largely by the owner's organisation. These objectives may, in the short term, be partially conflicting. If there are qualified local consultants capable of doing all or part of the work, the Project Director should encourage their use as part of his general interest in developing local capabilities.

If the necessary expertise is not available locally, and foreign consultants are needed, the Project Director should be interested in exploring opportunities for on-the-job training of staff through appropriate provisions in the consultants' contract. On the other hand, the Project Director should not endorse a strategy of developing local capabilities that would be at the expense of the quality of the end product. The consultant should be selected after the owner has suggested a joint venture with another group with complementary experience, on the basis of the experience of the people available and their knowledge of the particular geographical areas in which the project is being developed (Harding 1978).

A consultant's activities could include the design services only and another consultant could be selected for construction supervision. However, it is not unknown for the owner to select a consultant for all phases of the project including the construction phase. This will be determined according to the findings of the Base Line Study (I) and to whether all the elements of the project including funding, site availability, and schedule are clearly and properly defined.

The Project Director team will prepare the scope of work to be carried out by the A/E. The request for proposals (R.F.P.) will be based on the scope of work developed from Base Line Study (I) and the Project Organisation Plan prepared by the Project Director and his team. The Project Organisation Plan develops a strategy for organising personnel to carry out the project objectives. It delineates the levels of management, various project functions and personnel positions responsible for each function. This should include the owner's and consultants' personnel working together as one team with the Project Director as the team leader.

A combined owner and consultant staff will be assigned to manage and administer the project. The staff will be located in four bases:

1. Central Office. The Project Director and the Consultants' Executive Director should have offices in the Central Office together with a full staff of administrators, architects, engineers and support personnel. Policy and guidelines for design and site operations are established by this Central Office. Project control systems for scheduling, project accounting and quality control are also administered here as well as all variation orders, contract modifications, claims, disputes and requests for substitutions.

2. Site Office. A Site Director should be appointed as the personal representative of the owner, as well as a Resident Engineer as the representative of the Consultant on the site. The Site Director and the Resident Engineer, as the owner's representatives at the site, should at all times act together as a single entity with dual responsibility.

11

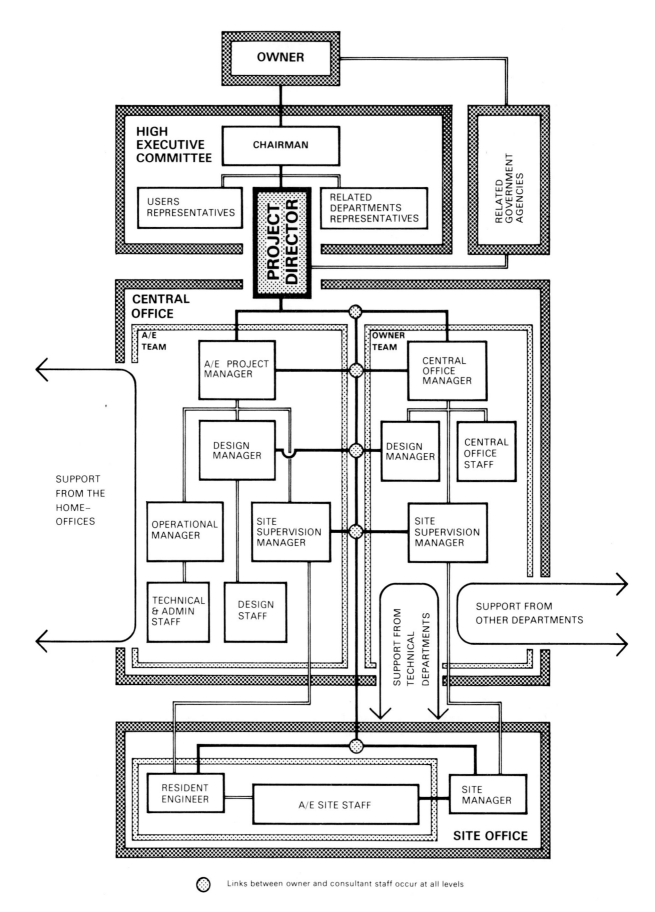

OWNER

HIGH EXECUTIVE COMMITTEE

CHAIRMAN

USERS REPRESENTATIVES

PROJECT DIRECTOR

RELATED DEPARTMENTS REPRESENTATIVES

RELATED GOVERNMENT AGENCIES

CENTRAL OFFICE

A/E TEAM

A/E PROJECT MANAGER

OWNER TEAM

CENTRAL OFFICE MANAGER

DESIGN MANAGER

DESIGN MANAGER

CENTRAL OFFICE STAFF

SUPPORT FROM THE HOME-OFFICES

OPERATIONAL MANAGER

SITE SUPERVISION MANAGER

SITE SUPERVISION MANAGER

TECHNICAL & ADMIN STAFF

DESIGN STAFF

SUPPORT FROM TECHNICAL DEPARTMENTS

SUPPORT FROM OTHER DEPARTMENTS

RESIDENT ENGINEER

A/E SITE STAFF

SITE MANAGER

SITE OFFICE

Links between owner and consultant staff occur at all levels

Fig.3: PROJECT ORGANISATION STRUCTURE.

The Engineer reports to the Central Office. Personnel in the Site Office deal directly with the Construction Contractor and conduct technical inspections and administration in accordance with the contract documents. The Site Office staff should be authorised to inspect working drawings and samples, to issue site instructions to the Contractor, and to prepare Pending Variation Orders for subsequent review by the Central Office.

3. Consultants' Home Offices. The design, production and technical support will come from all consultants' home offices. Support staff as needed will be available for call to the Central Office for technical assistance.

4. Owner's Technical Department. Assistance should be available from the owner's organisation and its divisions who will occupy the completed projects.

The project staff will include the following components:

1. Executive. The Executive includes the Project Director and the Consultant Executive Director.

2. Design Group. This includes the owner's Director of Design and the Consultants' Design Manager. These will work together and report to the Project Director and the Consultant Executive Director.

There will be a full Design Team at the Central Office. The Design Team should include all expertise necessary for the stages of planning, conceptional and schematic design, and would include planners, architects, engineers, quantity surveyors. The Design Team will be supported by home office teams for technical assistance and production.

3. The Operations Group. This includes the Director of Operations, Technical and Administrative Support Departments housed in the Central Office for both the owner and the Consultant. The owner's Director of Operations reports to the Project Director with responsibility for all activities assigned to the Operations Group. Activities assigned would include:

(a) *Contract administration.* The administration of construction contracts is a major responsibility of the Operations Group. This department considers the form and requirement of the construction contracts, the methods and procedures, reports and submission forms, as well as the functions of the project personnel at the Site and Central Office levels. All of these functions should be detailed in a Construction Administration Manual.

(b) *Technical support.* This is given by a large professional staff who provide technical advice and interpretation of architectural and engineering construction documents. Design services are also provided by this department whenever variations or additions to existing work are required. This group acts as principal adviser to the site offices with regard to the technical needs of the project.

(c) *Electronic data processing.* A complete computer system and operating staff exist to assist technical personnel in various aspects of their management responsibilities. The system should be capable of performing project scheduling, estimating, decision tracking, equipment inventory, and of providing a means for general management review. The E.D.P. should be installed in the Central Office to ensure reliability, productivity, quality control and maximum security of the project's sensitive data.

(d) *Office administration.* The Office Administrator is responsible to the Director of Operations for administrative support in the Central Office including: co-ordination of secretaries, drivers, translators, office attenders, purchase of office supplies, maintenance of office equipment, office security, maintenance of first aid and emergency equipment and operation of print shop.

(e) *Personnel management* It is essential to develop, maintain and manage a complete personnel programme to provide top quality professional and support staff to carry out Central and Site Office operations. The Personnel Department primarily recruits and mobilises employees to meet the manpower requirements of the project. Complete personnel files should be maintained which contain job descriptions, resumés, service contracts, customs information, time sheets and emergency contacts. This department also provides advice to staff regarding matters of annual leave, transportation, housing, dependants' education, medical and insurance protection.

(f) *Training and research.* Throughout the life of the project, the Training Department is required to develop and implement training programmes to enhance the knowledge and skills of the owner's staff. This division must also collect and disseminate technical reference material as required by Central and Site staff. Current professional literature should be collected in the Central Library for this purpose. This department should also be called upon to write and maintain procedure manuals which are useful to the project. A visual library should be developed with records of important meetings and presentations. Video films of actual site construction should be made on a regular basis.

4. The Site Supervision Group. The site supervision group includes the owner's Director of Site Supervision and the Consultant Resident Engineer and also other staff assigned to site supervision operations. The Director of Site Supervision reports directly to the Executive Director and is responsible for ensuring that the construction contractors carry out the work in accordance with the contract documents. Under the concept of dual responsibility, the owner should appoint a Site Director.

Personnel in the Site Office deal directly with the construction contractor, and carry out technical inspections and administrative functions needed to ensure that the work is performed properly. The Engineer reports to the Director of Site Supervision (Ministry of Interior 1982).

3
INCEPTION

Each project participant's involvement in the project may occur at different points in the programme. It may commence for the contractor when he sees a tender advertisement. It may start for a user when he needs more space. A government executive may be asked to put together an organisation to develop a project. The base line study is a way of describing the initial project for the user, the owner, or especially the government executives whose task is to develop the project. The tasks at inception are (see Fig. 4 overleaf):

1. Prepare a statement of owner/user needs (Base Line Study I)
2. Develop a project delivery method (Base Line Study II) or programme
3. Acquire funding (Funding Acquisition)
4. Acquire design professionals (Tender for Design Consultants)
5. Acquire a site for the project (Site Acquisition)

This book may be regarded as a base line study itself in that it deals with each element of the project delivery process and stresses the most important aspects of the base line study. Utilising the base line study approach, government executives ensure an efficient process in the project programme.

It can be demonstrated in Saudi Arabia, with such examples as Jeddah Airport, Ministry of Defence projects of the Corps of Engineers and the relative costs for the industrial 'new towns' of Yanbu and Jubail, that giving external consultants a freedom in determining the needs of the Kingdom in large scale projects has been a beneficial course of action. Individual ministries must develop sufficient expertise themselves to be able to control and regulate the planning stages of large scale projects from infancy.

However, it has become apparent from past experience that the Kingdom can no longer assume a minor role in the planning process of large scale projects. It is understandable, given the events that effected the Kingdom's growth over the last decade, that control over the sophisticated planning process of large scale projects was relinquished to 'export consultants'. The Kingdom at that time did not have the technological base nor experienced national experts who exist today. However, as the Kingdom enters into a new five-year plan it is essential to review the experience of the last fifteen years and now assert a leading role in the administration of large scale projects. The detailed development of elements of the base line study presented subsequently is intended as a future guide for control and management of a project. It can serve as a starting point for the project itself as well as providing assistance to those ministries now developing technical departments.

The base line study

One of the primary tasks in the project delivery system is to have the 'overall goals, objectives and constraints of the project identified, documented and agreed by all key participants in the development process' (Jacques 1976). The author identifies this process as a 'base line study' which is derived from the Arabic phrase 'el-derrsat el-awalyah' meaning starting point. It is an assembly of all elements necessary for the design process. The purpose of the base line study is to encapsulate the information necessary for the Architect to make design decisions. It also serves the Project Manager in decision making regarding programme and cost control as well as management process and scheduling. The base line study regulates the whole programming process in that it stipulates what information is necessary and relevant to the design of a particular project at any particular point in time. (In the private sector, this type of investigation and programming is done in the form of a feasibility study.)

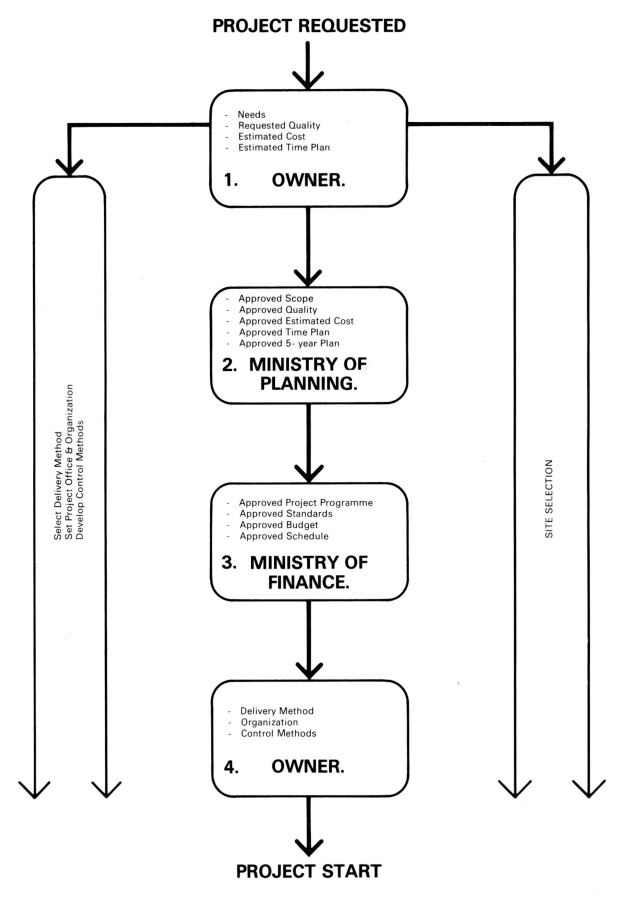

PROJECT REQUESTED

- Needs
- Requested Quality
- Estimated Cost
- Estimated Time Plan

1.   OWNER.

- Approved Scope
- Approved Quality
- Approved Estimated Cost
- Approved Time Plan
- Approved 5- year Plan

2. MINISTRY OF PLANNING.

- Approved Project Programme
- Approved Standards
- Approved Budget
- Approved Schedule

3. MINISTRY OF FINANCE.

- Delivery Method
- Organization
- Control Methods

4.   OWNER.

Select Delivery Method
Set Project Office & Organization
Develop Control Methods

SITE SELECTION

PROJECT START

Fig.4: INCEPTION PROCESS.

16

The base line study is a process that also requires the programmer to approach and study the whole problem of project delivery applied to a particular owner's needs and requirements. It is not the base line study itself, however, which will evolve into the solution of the problem. On the contrary, what it will do is give any agency a clearer idea concerning the problem and thus give a clear direction towards a possible solution.

Developing a base line study achieves, therefore, three major objectives:

1. it gives the initiating organisation a vehicle with which to determine the real needs of the problem and thus to develop a better understanding of the problem;

2. it is a tool that can provide the basis for justification of the project and subsequent funding; and finally,

3. it can be used to determine the process, methodology and tools of the project delivery system.

The elements of a base line study

In preparing a base line study, a primary function is to develop a methodology that will organise all phases of the project. Basically, each step of the base line study strives to clarify an objective or to identify a goal. Once analysed and translated into real goals and objectives, each element sets the parameters for the rest of the base line study.

The base line study is a two-phase process. It is a linear process, in that the first phase needs completion prior to entering the second one. Within each phase the process may become a heuristic one in that the information which is sought leads to a different sequence of steps. The first phase is a search and definition of the goals and objectives of the owner, and the second phase is a determination of how the goals and objectives will achieve a special physical reality In each phase, however, the overall parameters of the search and analysis are set by the preliminary problem statement as it is explored in relation to the determinants of Scope, Schedule, Quality and Cost. The key elements of the base line study are as follows:

Base Line Study I – Statement of Owner Needs

    1. Problem definition
    (a) problem statement
    (b) search for information
    (c) data collection
    (d) classify information
    2. Interpretation
    (a) scope
    (b) cost
    (c) time
    (d) quality

    3. Project goals
    (a) scope
    (b) cost
    (c) time
    (d) quality

Base Line Study II – Project Delivery Method

    1. Method of delivery
    (a) project process
    (b) design, construction, operation and maintenance requirements
    2. Management approach
    (a) project development team organisation
    (b) project management
    3. Method of project management control
    (a) project process
    (b) management plan
    (c) project management control systems
    4. Project implementation plan
    (a) method of project delivery
    (b) consultant agreement guidelines
    (c) architectural programme
    (d) operating procedures
    5. Preliminary funding proposal
    (a) design requirements
    (b) design phase deliverable requirements
    (c) tender phase requirements
    (d) construction phase preliminary requirements
    (e) operations and maintenance phase preliminary requirements.

Base Line Study I

Before the commencement of any project preparatory work has to be carried out; the true importance of this is often not recognised by many project managers. This document will include requirements of the user, will give a general idea about the project, will show the scope needed to satisfy the goals and the cost of it in terms of money and impact on the country's overall development.

It is the basic tool for negotiation with related agencies such as the Ministry of Finance and the Ministry of Planning. The study will be revised and updated once agreement is arrived at with the related Agencies and has a proven success record in Saudi Arabia.

Most projects fail at ministry level because insufficient evidence is presented by the proposers of projects as to their necessity or financial support. These failures are often attributed to lack of co-operation at a bureaucratic level.

1. Problem definition

(a) *Problem statement*

Through the process of a Base Line Study I, the programmer sets out to organise the information which he gathers when analysing the problem. The programmer therefore establishes a preliminary problem statement. Based on this statement, the programmer searches and collects data. It is helpful, in formulating the preliminary problem statement, to develop it in relation to the primary determinants of any design problem: scope, schedule, quality and cost. This will make the preliminary problem statement comprehensive enough to allow a broad search and at an early stage force the programmer to examine the whole problem. As W. Pena states, in his book, *Problem Seeking*:

> There is the need to amplify, to view the whole problem; but there is also a need to abstract. You amplify and then narrow down; you seek the ramifications of the information gathered; then you turn around to determine the bare implications. It's a continual process. You must be able to see the trees and the forest – not both at once but consecutively, from two different points of view. (Pena 1977.)

As one proceeds through the process of data collection and analysis and begins to learn more about the problem, the preliminary problem statement will eventually become the final problem statement.

There exists a cycle which will occur within the step by step process of the study itself. It may be defined as an iterative process (Palmer 1981). It allows the programmer to refine objectives, needs or statements based on analysis and discard irrelevant information. A programmer, for example, will start collecting data and begin to analyse the project to determine the actual scope. During this analysis he may discover a need that was not realised when the problem statement was initially formalised. He will then go back and redefine the problem statement and then, as a result, begin to re-analyse his data base in relation to the refined statement.

During the problem definition period, it is a good idea to set the range and limits of the scope of the study. On developing a base line study for a large scale project, the programmer should look at what factors are involved, especially in terms of time, and make sure that he sets limits appropriate to both the study and project needs. In many cases, there will be determinants outside the programme's control which will themselves determine its limits. In government agencies, the development of a project is usually part of a whole development plan for the country, which in turn will dictate a larger time element within which the project elements must conform. In other cases, the need to co-ordinate government funding and budgeting requirements will set appropriate limits. The programmer should identify these constraints so that he may focus his energies in those areas that will be the most productive.

(b) *The search for information*

The process of a Base Line Study begins with a search for more information and its purpose is to expand and fill gaps in the problem statement issued by the owner/user. The sources of information for search purposes will vary according to the type and nature of the project, but may include:

- Owner user
- Owner/user consultants
- Proposed site
- Neighbourhood
- Codes and 'Standards' publications
- Technical publications
- Similar existing or proposed facilities
- Special consultants
- Relevant authorities and ministries
- In-house expertise and data

Search methods are not definitive. The process itself will be dynamic in that one revealed source of information will often identify a further source to be explored. Often, the most logical first source will be that from which the initial enquiry has come – the owner/user. This source will establish his:

- Usefulness as a source of information;
- Knowledge of any existing data bases;
- Knowledge of any existing examples which may illustrate his requirements.

The dynamic search process will have begun.

(c) *Data collection*

Data collection is the process of gathering information which then forms the database for determining what is needed.

The process is essential to ensure that information regarding scope, quality, time and cost is gathered in a way which provides easy access for the study team. It is the physical result of the information search.

The content of the gathered database will also define the adequacy of the search operation. If gaps in the database information are discovered at this or any subsequent stages of the process, further search will be necessary for data which will fill these gaps. The process of determining what is needed is therefore not a linear progression through steps 1–6 but will be a cyclic process back and forth to ensure completeness of information required.

The sources of data collection will be those based on the initial and subsequent searches. (See (b) *The search for information* above.) Within these sources it is important to identify the best 'sub-sources' for collecting useful data – a particular department within an organisation or a particular individual for a particular subject. For example, an owner body may have several decision-makers within its committee or organisation. The eventual user may be a Decision maker A regarding scope and the owner may be Decision maker B – on cost. There may be one source of decision making, or there may be several, but these need to be identified as soon as possible. The quality and timing aspects of the study will in part be determined by obtaining data from sources A and B since both aspects are inter-related in scope and cost.

Many methods and techniques can be employed in data collection and data exists in different types. The method of data collection needs to be that which is best suited to the type of data to be retrieved.

Methods of data collection include:

- interviews (questioning, discussion)
- consultations (questioning, discussing, taking advice)
- surveys (collecting opinions; attitudes; finding out what exists)
- research (background understanding) – collecting facts, conceptual trends; projection trends; self education to fill gaps in knowledge
- visits (combination of all the above and comparative tool for appraisal).

In the search for data, a primary source would be the eventual user groups and related public or private agencies. In surveying these groups a multitude of data collection techniques can be used. Each technique is suited to obtaining a certain type of information. 'Surveys are used primarily for collecting opinions and measuring attitudes. Data logs and standardised forms seek to document factual, descriptive information. Interviews and questionnaires can be used to obtain both descriptive and evaluative data.' (Palmer 1981.) Another major area of data collection would be the collection of background data. The most important activity to occur here is a records search for any data which could be related or relevant to the subject project. For example, the author, some years ago, developed a base line study to determine the housing needs for existing and future staff of the Ministry of Interior. During the data collection phase, research into existing documentation revealed a study by the Department of Municipalities for the planning of settlements throughout the Kingdom. This source produced a wealth of information about demographic trends, social-economic information and population movement projections that saved much research time in developing the base line study for the Ministry. The actual process which uncovered the existence of this document was a series of unstructured interviews with other government and local agencies who, it was felt, would be directly affected by the implementation of the project. This was in itself another technique for collecting background data (the unstructured interview), which is a useful tool for establishing the scope and nature of a project. However, a description of all the various techniques and tools available in data collection would prove too lengthy for the purposes of this discussion.

(d) *Classification of information*

The base line study programming process continues into its next phase by classifying information gathered during the data collection phase. The classification of information is the arrangement of data into groups or classes. Classification simplifies the problem and establishes a systematic approach. Data is classified into the groups already determined, i.e.

Scope
Quality
Schedule
Cost

It will be remembered that one of the goals of a base line study was to 'define the limits of: scope; quality; schedule and cost.'

Consisting of four primary elements, this framework leads the programmer through interpretation to goals.

1. SCOPE	(a) *Quantitative*	Size; activity scale; people; area; schedule of accommodation; schedule of staff;
	(b) *Qualitative*	Performance: people; site; environment; relationships;
2. QUALITY	(a) *Quantitative*	Life expectancy; standards/codes; periods of fire resistance;
	(b) *Qualitative*	Specification; spaces; standards/codes;
3. SCHEDULE	(a) *Quantitative*	Programme/schedule; occupancy date
	(b) *Qualitative*	Time perspective; design development and construction time;
4. COST	(a) *Quantitative*	Initial costs; unit costs; area; volume; net; gross; return;
	(b) *Qualitative*	O + M; life cycle costs.

This gives eight numbered groups of data collection. At this stage, data may be entered under more than one category heading. This is relevant to the overall programme process in that it highlights areas of conflict or applies emphasis (in the case of cross reference agreement) to an item of data already double-entered. Areas of conflict can be identified and tested.

The data sorting process has two main objectives:

1. Refinement by simplification

2. Elimination of irrelevant information

The sorting process focuses the programmer's attention on the essentials which will form the content of the problem statement. It keeps the major aspects of information and abstracts irrelevant data. Only essential and relevant information is carried forward to the interpretation stage.

The results of the sorting process are:

1. A data base which is relevant and useful.

2. Essential preparatory work achieved prior to entering the interpretation stage;

3. Identification of any gaps in information;

4. Content which is conflicting and can be resolved through further interaction with the owner/user;

5. A 'background data' file compiled with all abstracted information.

2. Interpretation

The main objective of this task is to determine project needs in relation to potential and limitations. It should be remembered that each problem has inherent potentials and limitations built into it: thus, every problem statement implies a solution. If the programmer can identify the potentials and limitations, he is progressing towards the solution of the problem. Take, for example, the hypothetical problem of a need to build a first-class hotel in the city of Los Angeles prior to the opening of the 1984 Olympic Games. What are the potentials and what are the limitations of that problem? The answer to that question might look something like this:

Hotel Project: Potentials and limitations

Problem statement:

• To build a first-class hotel in the city of Los Angeles to be ready for occupancy in time for the start of the 1984 Olympic Games.

Problem potentials:

• Los Angeles is a good construction market for labour and materials.

• The Los Angeles area imposes no limits on the technology of the project.

• Private funding is available for this building type.

• The city has the market to support another first-class hotel.

• The 1984 Olympics will provide the necessary boost in revenue to allow the hotel to make a profit in the first year of operation.

Problem limitations:

• The design and construction time is short, approximately one year.

• The only suitable site close enough to the Games

has a height restriction.

• There is the need to acquire additional land in order to satisfy the economics of city parking regulations.

• Financing is available, but has limitations.

• Profitability of hotel operations during post-Games years.

These statements act as a checklist, to which the collected and analysed data is applied, and help provide the framework of a solution. Each statement defines an area of potential to be explored, or a limitation which must be dealt with during the programming and design phases of the project delivery system. What the setting of the potentials and limitations of the problem does, is set a range of possibilities as to how the project can be achieved. They set the parameters of the solution, which helps to overcome part of the problem.

Every problem has a definable scope. Even a project of indefinable duration has scope, in its very lack of definition. The problem of a project of indefinite duration is a good example of how each of the primary determinants are inter-related and affected by the others. Scope is affected by time in a project of indefinite duration. Quality in this respect can be isolated to a degree, but cost cannot: as time increases so does cost. If cost were the factor being defined, then quality could no longer be isolated, for as costs are reduced, either quality, scope or time would have to diminish. This is the crux of the interpretation process: the inter-relationships and effects of the primary determinants against the project needs, within the parameters of the limitations and potentials of the problem.

The interpretation phase is the 'heart' of the base line study, for it is here that project needs are determined against the primary determinants of scope, schedule, quality and cost and within the parameters of the potentials and limitations of the problem. What the programmer is trying to achieve in this process of interpretation, is to analyse the project needs in order to arrive at a set of goals which are realistic and satisfy the needs of the problem.

This process can be simplified by applying a simple equation of logic. What the programmer is really doing, when he first attempts to define the needs of the problem, is to establish the 'ideal goals' that would satisfy the problem. Therefore, 'needs' equals 'ideal goals', or

$$N = IG$$

where N is needs and IG is goals. Ideal goals signify the ideal conditions that would allow the 'ideal' solution to the problem, but are seldom achieved. In the case of a large scale project, they are seldom achieved: due to the potential and limitations of the problem, and outside determinants of scope, schedule, quality and cost, 'ideal goals' are interpreted into 'real goals'. These are goals capable of achievement in relationship to the determinants and objectives of the problem.

This means that real goals are a function of ideal goals as affected by the potentials and limitations of the problem and the primary determinants of scope, schedule, quality, and cost as they relate to needs, or

$$RG = f(P/L + S,S,Q,C)IG$$

where RG is real goals, P/L is the potential/limitations of the problem, S,S,Q,C, are the primary determinants of the project, and IG is ideal goals. This is the essence of the factors that are brought to bear in the interpretation of 'ideal goals' into 'real goals'. The more thorough the steps of data collection and analysis; the more clearly the potential and limitations of the problem have been defined; the more precise the problem statement; the more accurate will be the interpretation of 'ideal goals into 'real goals'. The process is not quite as simple as it seems here, because all of the determinants are inter-related and a change in one affects all the others. This can be simplified in the following formula of how cost is related to the other primary determinants:

$$C = f(S,S,Q)$$

Where C is the Cost and S,S,Q are the other three primary determinants. It can be said that the other three relate in the same way.

3. Project goals

This stage is the final result of Base Line Study I. Its main task is to develop the objectives of the project, evaluate the various alternatives, and then to determine the project goals.

In his book, *The Architect's Guide to Facility Programming*, Palmer indicates five components which can be considered as the principal ones of many programmes developed recently. With the exception of item 5, which has to do with energy requirements (though the author is convinced that this is not a principal component of a programming or base line effort, and is dependent on whether energy is an issue to be considered; this will vary from one part of the world to another), there is a direct identification of the four primary determinants described in previous sections of this book:

1. analysis of functional and/or activity systems to be accommodated (determining the need in terms of scope);
2. determination of the kinds and criteria of space corresponding to functional needs (determining the need in terms of quality);

3. an estimate or analysis of the cost implications of physical requirements (determining the need in terms of cost);
4. establishment of schedules for carrying out the programming, designing, construction and, sometimes, occupation of the time);
5. Analysis of the requirements and implications of energy use.

These primary determinants are the checklist to which ideal goals are applied in the determination of real goals. But, what is the process of determining scope, schedule, quality and cost? The answer is difficult and entails research, data collection and analysis if the question is to be answered properly. Take, for example, the housing project on a large number of sites used as a model throughout this book. The objective of determining the scope of the project would be to answer these questions:

1. What quantity of housing is required?
2. What space/accommodation standards should there be?
3. What density of housing?
4. What occupancy size(s) of housing?
5. What size of neighbourhood?
6. What type of neighbourhood?
7. Where should the sites be?
8. What need is there for accessibility of the sites?
9. What support/service systems/facilities are required?
10. Are there any specific needs of organisation groups above normal housing needs?

In dealing with the schedule, the objective would be to answer the following questions:

1. When is the housing required?
3. How many are required in what time increments?
3. What types are required and when?
4. When will certain areas require housing?
5. How does this affect project delivery?

In determining the quality, the objective would be to answer the following:

1. What standard of space/accommodation is required?
2. What standard of construction?
3. How will standards differ for different organisation groups?
4. What type of amenities are required for each type of neighbourhood?

Finally, in determining the need in terms of cost, the following questions would have to be answered:

1. What is affordable?
2. When will it be affordable?
3. What should be the unit cost for each type of unit?

4. What other costs will be encountered and when?

It is intentional that the determination of need in terms of cost is the last item above. Scope, schedule and quality are all factors that will affect cost. However, it is necessary to establish answers for the other three before costs can be properly analysed. Again, it may be helpful to remind the reader that during this process criterion cycles will occur and, as analysis progresses, needs will be redefined.

The lists of questions under each item above are not intended to be comprehensive and are particular to that type of housing project. As the problem changes so will the questions. The questions above seem to be a simple step in logic but what may be misleading is that there is no way of checking whether such a list identifies all the elements of the problem unless a framework is developed in which the determinants are tested against the parameters of the problem.

Base Line Study II

Project approach

The main purposes of Base Line Study II are to select the system needed and to make the project goal a reality. The approach is essentially intuitive and should be started from scratch.

The Project Director should ask all the necessary questions, to which the answers determine the approach. The important general questions are:

Where am I going?
Where am I now?
What is the total need?
How do I get the things I do not have?
How do I manage to get it done?

1. Question? Where am I going?

Answer: To complete a building project and to turn it over to a user in a way that meets the user's needs in terms of function, operation and maintenance.

2. Question? Where am I now?

Answer: I represent the owner's project development authority, with only a general building programme, with an idea of the scope of the project, the quality and standard desired, the time requested to complete the project and one of many sites which could suit the project.

3. What is the total project need?

Answer: Choose from items A–X.
A. Project Managing Entity.
B. Building Programme (Schedule of Areas).
C. Project Design Data (including constraints)
D. Project Requirements: Scope, Quality, Time, Cost

E. Project Delivery Method
F. Funds for the Site
G. Site Acquisition
H. Selection of Site
I. Funds for Design Consultants
J. Design Consultant
K. Expediting/Controlling Design and Construction Documents
L. Design and Construction Documents
M. Funds for Construction
N. Contractor Acquisition
O. Contractor
P. Expediting/Controlling Construction
Q. Construction
R. Project Hand-over/Turn-over to Organisation
S. Project Occupants' Satisfaction
T. Funds for Operations and Maintenance
U. O & M Organisation Acquisition
V. O & M Organisation
W. Operations and Maintenance (Technical and Non-Technical)
X. Feedback for Future Use.

4. Question: How do I get things I do not have?

Answer: There are alternative choices which must be analysed and the best possible alternative for items A–X be chosen.

5. Question: How do I manage to get it done?

Answer: This answer is, in essence, the Base Line Study Part II, which is:
- Management Approach
- Delivery Method
- Method of Control

The Base Line Study II is itself part of the project need that falls between other project needs (see chart opposite).

Analysing Project Needs

A. *User's agency involvement*
The individual or group which utilises the project must be considered in terms of the extent of their involvement in the project process, both in terms of capability and information requirements. Considerations are:

- Facility utilization
- Operation and maintenance
- Construction management process
- Selecting contractors
- Design input
- Site acquisition
- Selecting design consultants
- Acquiring Funding
- Managing the delivery process
- Programming input
- Extent of organisation involvement and capability.

It is necessary to understand the above considera-

Base Line Study Part I

Project Requirements –
Scope, Cost, Time, Quality

Base Line Study Part II

Project Delivery Process
1. Management Approach:
   Management Entity
   Policy
   Procedures
   Management and production resources
   Constraints
2. Delivery Method:
   Design, Construction and O & M
      Delivery Method
   Project process
3. Control Methods:
   Scope
   Cost
   Time
   Quality

Funding Proposal (may have started prior to BSII)
Design Consultant Selection
Site Acquisition (may have started prior to BSII)

---

tions in terms of determining the ultimate efficient use of the project. Too often the organisation has preconceived ideas about the project process without taking into account the requirements for the finished product. For example, how long should it last? What are availability factors for areas used by tenants? Who provides social services? There must be an identification of special needs, to determine which consultants are required to provide for those needs.

1. Utilizing the facility. Will the user agency be involved with the project for its full life, or for a period of time before passing the building to another user, or acting in the capacity of landlord over tenants using the facility?
2. Operating the facility. Who will be responsible for operating the services – the owner or the user? Or would it be a shared responsibility between those who operate a service?
3. Operations and maintenance. Who will decide the way in which the project is to be maintained from the beginning and identify the requirements?
4. Construction management. Will the user's agency take part in the construction manage-

ment process which gives it a more intricate knowledge of the facility?
5. Selecting the contractor. Is the user's agency qualified to make choices or decisions in selecting the contractor?
6. Design process. How can the user's agency make the best contribution to the design process? Consider each phase of design, including:
- programme
- outline design
- design development
- construction contract documents

7. Site acquisition. Can site acquisition best be done by the user's agency?
8. Selection of design consultants. Is the user's agency qualified to make choices or decisions in the selection of the design consultant?
9. Project delivery process. Can the user's agency manage to execute part or all of the project delivery process?
10. Acquiring funding. What role will the user's agency have in acquiring project funding?
11. Programming. To what extent can the user's agency provide project programming data?
12. User involvement. What does the user's agency need in regard to the full extent of assistance?

It is essential to utilise the opinions of the future occupants of the project but to recognise their limitations in specialised knowledge.

B. *Project managing entity*

Five alternatives are usually considered for managing multi-disciplinary groups for large scale projects:

1. Manage within the existing department
2. Establish an extension of the department
3. Manage through another government agency
4. Manage with services assistance from a managing consultant
5. Hire a managing consultant to manage.

This study proposes a mechanism for the execution of a project without a managing consultant taking full control. Managing the project is a process of knowing what choices exist and making the best possible choice. It is essential, therefore, to utilise the most appropriate management tool, which will generate alternatives for consideration.

C. *Building programme*

Base Line Study I produces a building programme in a form which can be utilised by design consultants. The responsibility during BSII is to define the flexibility and constraints of the programme in the design process. If planning may be altered, depending on an alternative design approach, then this should be made clear at the inception of the design.

D. *Project design data*

Most design data is considered in completing BSI. Design data is completed and categorised in BSII to ensure that both the designer and project manager have the information for the design and for understanding the constraints which govern the project.

Two types of project design data are relevant:

● Data to support design
● Codes and standards

Data to support design. The organisation can give the best information to make the design function successfully if the organisation is asked the correct questions. The organisation can give design input if shown an imaginary walk through every metre of the site or building. This may be tedious but serves as a useful checklist of items from BSI which can be given to the designer. This should be followed up in the programme analysis and schematic (outline) design phase with further questions concerning, for example:

● Frequency and size of service vehicles
● Underground shelter approaches
● Water table problems in surrounding areas

Codes, regulations and standards. Codes from other countries can provide a safeguard required for most components of the project. Some data from codes and regulations might be used for reference purposes only, if not directly applicable in Saudi Arabia. For example, American standards may be specified by an American designer for one part of the project, and in another part a European engineer may apply a European standard, but if the contractor uses Saudi building products these standards may not be applicable.

It is better to be specific about which code or standard is practical for each part of the project. Requiring performance standards is the best safeguard in a project where it may not be practical to determine the component supplier at an early stage in the project.

It is important to make the designer aware of and concerned for realistic codes, regulations and standards. Experience of standards applied to local conditions may vary from one international firm to another. Many firms may not have adequate quality control or recent experience to ensure that codes and standards are applicable to

certain areas or types of projects. Multi-faceted projects often run the risk of a firm having minimum experience in just one particular facet of a project. This can be perilous in complex building types such as hospitals or medical clinics.

Each Saudi Arabian Ministry concerned with the design or construction of a project has now developed a series of standards and a knowledge of which international codes, regulations and standards may be appropriate.

Correct local municipal codes and standards can be acquired during BSI and given to the designer to avoid waste of time and omissions occurring from his not being acquainted with them.

Site acquisition

In the early stages of a project, initial meetings with the user's representative should take place to establish a project site. Normally the user's agency will be responsible for finding the site. In this case, the project team's responsibility will be limited to evaluating the available site, and determining its limitations and potential; then to deciding either to go ahead with the available site or recommend that another site be located. Often, cases arise where either the site is not adequate or it cannot be provided by the user's agency. In this case the site could be provided from one of the following sources:

1. Other sites owned by the user's agency should be checked to determine whether they will meet the requirements of the project.

2. Sites owned by government agencies such as the Ministry of Finance or the Ministry of Municipality.

3. Sites owned by the private sectors.

The first step for the project team is to determine the required area for the project (supported by the base line study), the preferable location, and the characteristics of the needed site. Then the Project Director should prepare a form which should be signed by the highest possible person in the agency, to send out to every government agency which, he assumes, owns land in the required area. In case of a failure, the Director should start preparing for a hard battle with the Ministry of Finance to provide the required funds to buy the site from the private sector.

4
PROJECT CONTROL
AND COMMUNICATION

Methods of control in project management

For project management to be effective, there must be a form of project control which will allow the project manager to monitor the progress of the project as it unfolds step by step, providing him with the tools to analyse its progress and also providing the mechanism for corrective action when problems arise.

Developing an effective project control system is contingent on first producing a statement of what the goals and objectives of the project may be, not only in terms of quality, cost and time, but also in terms of the methods or approach desired for implementation. To develop a system of project control there should first be a clear indication of the scope of the project and the required schedule, and predetermined budget. When these parameters are established, they become the yardstick against which the actual schedule and costs can be measured.

A project control system is built upon the principle that an estimate and project plan is prepared early in the life of a project. Cost and progress are continually compared to the planned budget and schedule (this must occur during all phases); costs and progress are reported, forecast and controlled. This element of control is a responsibility of management and not inherent in the system.

Control is achieved through the Project Manager's first establishing a set of procedures which allows him to monitor the performance of a project against a pre-determined plan; and secondly by his actually controlling the progress of work to ensure that adequate performance is maintained. Monitoring is carried out through a process of recordings and measurements of actual project activities and then by analysing them against the planned objectives. Control is achieved by an actual course of action taken to remedy an unfavourable performance. This is done by managerial effort or a 'plan of action' to rectify a delay or cost over-run and to bring the project back on schedule or within the budget. For example, in monitoring the design phase of a particular project, the project manager may analyse the man-hour time sheets (a form of measurement) and notice that the production of furniture layout drawings is taking 30% more time to complete than was originally planned. Through monitoring procedures which are set up by the project control system a problem area has been identified. Aware of the problem, the project manager can analyse it and demand corrective action. Four important criteria operate in the example above:

1. There was a system of monitoring already built into the project process (the project control system).
2. There was a predetermined plan with which to measure the progress of actual work (a base line study).
3. There was a method of reporting (the monitoring process) which was clear and simple enough for the project manager to understand.
4. The problem was identified before it reached a critical stage.

A project control system has two main objectives to achieve in order for it to be effective: (1) it must identify deviations from the predetermined plan; and (2) it must summarise data and assess trends to identify future problem areas early enough to allow management to apply remedial action before the problem reaches crisis proportions.

Monitoring performance

To be able to monitor the performance or progress of a project, a basis for comparison must be established before the process of monitoring can be performed. This could be applied equally to a teaching situation. A teacher has a predetermined set of goals in the education of his or her students. It is, in simple terms, to educate the student in a particular area of knowledge. The teacher prescribes a series of tests at intervals during the duration of the course. Those tests act as an indicator to the teacher of the actual progress each individual student is making in the achievement of the predetermined goals set for the course. This same philosophy can be applied in principle to a project control system. It is a way for the project manager to evaluate the progress of work against a predetermined plan. But, instead of prescribing tests, the project manager monitors by the accumulation of specific information. This type of feedback, as in educational test results, provides the project manager with an indicator of where more attention needs to be paid and where effort needs to be concentrated to achieve better results. This, then, describes the process of monitoring, which is an important part of any element of project control.

The specific information which a project manager needs to accumulate for a particular project in order to monitor the progress of work can vary. It is dependant on what specific part of the plan is designated as the most critical. For instance, if the project has a critical budget, then it is necessary to incorporate an element of control, and the specific information to monitor and control would be cost. On the other hand, if the programme of the project was important, then the specific information requiring monitoring would be progress of work. As in the development of the base line study, the most critical elements of a plan must be established and all these elements inter-relate. If cost is the critical factor, then there is a resultant effect on schedule, scope and quality. In addition to monitoring the most critical element of any pre-determined project/plan, the effect on the other elements must also be considered.

The development of a project control system for the total project delivery process requires the project manager to realise that in the delivery process there are different areas of project control. When developing a base line study, it is difficult to develop a method of monitoring and control of the work as the scope, schedule, quality and cost of the project has not yet been defined. That does not mean that the development of a project control system can be ignored at this early stage; on the contrary, it must be considered, but in a different manner.

The base line study

During the process of developing a base line study, the project manager must take an active role in helping to incorporate the project control system into the base line study itself. He should monitor the development of the project control system within the base line study to ensure that 'it will reflect the level of detail to actually control the level of detail within the project itself' (Havilard 1981). The project manager should also take note that as the project control system will be only one element in the management process of a project it should be designed to fit in with the project's overall system of quality, financial and programme management.

Monitoring during design phase. Monitoring the design phase is much the same as monitoring the whole project process. O. Goodall, in his course on project management, uses the following analogy:

> A piece of equipment is expected to meet the requirements laid down for it in a specification. So the specification may be considered to be the base line criteria. As the piece of equipment is operated and as the performance is measured, the measured values of actual performance may then be compared to those identified in the specification and any variances identified. (in Foxhall 1972).

The same can be said about the design process. The base line study sets the standard criteria on which the performance of the programming and schematic design efforts is based. As the project enters into the design development and construction document phase, the design has become fixed and development of relative details can be evaluated against a set of design control procedures and standards, according either to the requirements of quality set out in the base line study or to those developed by the firm doing the actual design.

Measuring performance during the design

One method of measuring performance is to have a basic system of determining whether the progress of work performed on day x at the cost of y is in line with the projected progress and cost. During the design process, an effective control system requires that three aspects of the work be monitored: schedule, budget and quality. The project manager should be able to measure, on any given day, the progress of work in relation to these aspects and then by comparison with the original plan determine whether performance was adequate or, if not, to adjust accordingly.

Monitoring schedule

A schedule can most effectively be monitored during the design process by a measurement of man-hours expended. With the scope of services determined, the project manager takes each task to be performed and works out the man-hour requirements needed. There are a number of different methods for identifying these. One such method is for the project manager to negotiate with each member of his team the time required to perform a specific task. The task man-hours are then gathered and plotted on an overall schedule. As the work progresses, each team member records the actual man-hours expended on the project. These are accumulated on a weekly or bi-weekly basis (the shorter the duration between reports, the more accurate the results), calculated and logged against the proposed schedule, giving the project manager a clear comparison of actual, against proposed, programme. Since this task is almost entirely quantitative, the use of computers can greatly speed the process. Every task has a schedule of time for its completion. With programme control, there are also the related factors of cost, quality and scope.

Project control system

The Organisation Director and Project Manager(s) should control a visual graphic and written display of information which will allow the measurement, evaluation and correction of projects, plans and tasks. These controls should also allow for decision making, problem solving and response to questions.

Project control reports are developed in the form of Project Manager reports and data which include:

Descriptive Information
  ● Scope of project reports
Controlling Information
  ● Cost reports
  ● Time schedule reports
  ● Quality assurance reports
  ● Problem reports
  ● Activity reports
Planning Information
  ● Planning reports
  ● Project information reports

Reporting should be done in a uniform format. Information in the reports will be limited to significant matters which affect major milestones and important project activities. The department's Directors will provide overall department project activity information and also specific project information from Project Managers. Each Project Manager will have established the levels of reports and types of information required, based on specific needs and the requirements of Department Directors both for department use and for the Organisation Director.

Reports which may be required by the Organisation Director may include:

Activity reports

● Activity tracking
(Dates, required action, persons involved, etc. for various items including task force directives, payments, change orders, milestone events, letters)
● Monthly reports
(Past and future activities, financial summaries, cost and schedule data)
● Situation reports
(Information reports on specific items such as site clearance, excavation, roof completion, site problems and solution)

Planning information

Project loading report:
● Department project loading
(Department project listing, type, size, department start and projected completion date)
● Department task loading
(Department special project related to external assigned project duties)
● Project control board up-date information
(Project manager reports specific changes to the project in terms of scope, cost, schedule, or quality level).

Descriptive information

Scope of project reports:
● Project description report
(Size, occupants, location, participants, cost, etc.)

Control Information

Cost reports:
● Cost and schedule summary (Also on time schedule reports)
(Cost of work performed, contracted, budgeted, comparison)
● Payment log
(Invoice – contracted and agreed)
● Variation order log
(Contracted amount, changes proposed and accepted)
● Cash flow comparison
(Projected, agreed, actual cash flow)
Programme reports:
● Cost and schedule summary (also in cost reports)
(Schedule – completion dates, projections, agreed, comparison)
● Time schedule overview
(Graph showing months ahead or behind schedule)
● Activity schedule
(Categories of activities by PHASE, required by consultants, contractors or company contracts)
Quality control reports:
● Quality control reports

(Comments by technical review teams on deliverables or comments by construction inspectors on work quality. Both are official reports for mailing to consultants, contractors or companies)

Problem analysis reports:
● Problem report
(Description, cause, alternative solutions, recommended solutions, action needed)

Project management and the communication process

It is only recently that the importance of project management in the success of any project has been realised. The Project Manager plays a central role in the successful completion of any project within the specified time and within the available resources. Because of this central role, the project manager is a key person in the communication process of a project. Being the central element, he has the ability to control, direct and motivate the other participants of the delivery process. Project management has evolved as a means of coping with the increasingly complex task of communication in both directions within a hierarchy as well as externally.

As projects became more complex over time, the project manager's role changed from the traditional mid-management one in the line of authority to one in an important central position. Fig. 3 (p. 12) graphically illustrates the project manager's position in relation to other key participants in the project delivery process. An Information and Communications Manual should be developed by the project manager as part of a continuous process. Important elements in this manual might be:

Methods of Communication

Letters
Memoranda
Telex
Facsimiles
Meetings
Meeting notes
Reports
Verbal communication (general)

Types of communication

Problems
Strategies
Task force directives
Instruction (how to perform a task)
Selling an idea
Clarification
Situation analysis
Argument
Refutation
Topical information
Research reports
Procedure

5
CONSULTANT SELECTION

The selection process

The initiating organisation seldom has the in-house capability for performing all programming, planning and design tasks required of large scale projects. In most cases, the initiating organisation will develop an RFP (Request For Proposal) and issue it to a selected group of consultants. The scale, and the related high financial investment, in large scale projects requires the development of a methodology for information gathering and evaluation which will ensure a comprehensive analysis of the firms eventually selected to tender for the project. The methodologies developed for A/E consultant selection by the author are described in the following pages and were utilised in most of the projects for which he was responsible in management terms. (See Fig. 5 overleaf.)

Selection methodologies

In order to ensure a proper procedure for selecting fully qualified design firms, an evaluation of existing pre-qualification guidelines established by international, governmental, financial and professional organisations should be undertaken. The procedures developed by these bodies may not always be totally applicable, however. Nevertheless, they will serve as a starting point and a reasonable check list. Organisations which have developed refined pre-qualification guidelines are as follows:

- Federation International des Ingenieurs-Conseils (FIDIC)
- International Bank for Reconstruction and Development (World Bank)
- Inter-American Development Bank (IADB)
- United Nations Development Program (UNDP)
- American Institute of Architects (AIA)
- American Institute of Planners (AIP)
- Royal Institute of British Architects (RIBA)
- Agency of International Development (AID)
- USA Housing and Urban Development (HUD)
- UK Department of the Environment (D of E).

The selection procedure outlined here was for a multi-phased housing project, to be carried out over a ten-year period, which involved the programming, master planning, design and construction management of 40,000 units of housing on 30 different sites throughout the kingdom. For the purpose of this example, US Government forms nos. 254 and 255 were selected as a standard submission for pre-qualification data by the various firms invited to tender in Saudi Arabia. It was realised when evaluation of the initial pre-qualification data was received that no one firm had the total capabilities to produce such a project. The final recommendations were to select pre-qualified consultants grouped together to form consortia or associations which would give them, as groups, the capabilities required to implement such a project. Once evaluation had been carried out on all pre-qualification data, a short list was developed of firms considered suitably qualified and interested. The criterion used here for evaluation was based solely on the professional capabilities of the firm with no bias towards any special requirements of the project. In other words, if the firm presented itself as strong in architectural design and planning, then it was so evaluated. The result was a short list of firms, with each firm having a designation of those areas it was strongest in, e.g. CRS (AE) – where A, being first, indicates primary strength in Architecture, and E secondary strength in Engineering; or DMJM (AEP) – A, Architecture as primary strength; E, Engineering as secondary strength; and P, Planning as tertiary strength).

The next step in the selection process was to set up screening interviews with all firms on the short list. These screening interviews were undertaken against a set of performance criteria that, this time, emphasised the special characteristics of the project. Prior to interviews being held, an evalua-

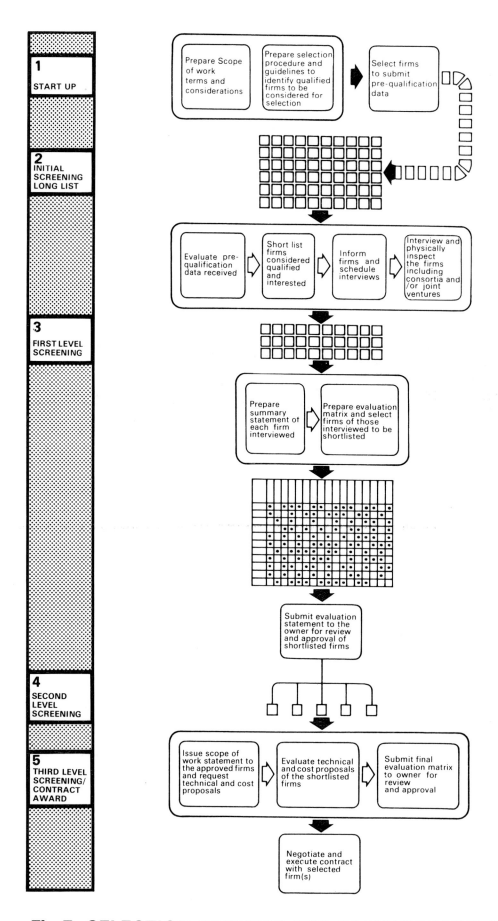

1	START UP
2	INITIAL SCREENING LONG LIST
3	FIRST LEVEL SCREENING
4	SECOND LEVEL SCREENING
5	THIRD LEVEL SCREENING/ CONTRACT AWARD

Prepare Scope of work terms and considerations

Prepare selection procedure and guidelines to identify qualified firms to be considered for selection

Select firms to submit pre-qualification data

Evaluate pre-qualification data received

Short list firms considered qualified and interested

Inform firms and schedule interviews

Interview and physically inspect the firms including consortia and /or joint ventures

Prepare summary statement of each firm interviewed

Prepare evaluation matrix and select firms of those interviewed to be shortlisted

Submit evaluation statement to the owner for review and approval of shortlisted firms

Issue scope of work statement to the approved firms and request technical and cost proposals

Evaluate technical and cost proposals of the shortlisted firms

Submit final evaluation matrix to owner for review and approval

Negotiate and execute contract with selected firm(s)

Fig.5: SELECTION PROCESS FOR CONSULTANTS SERVICES.

tion sheet was developed and each firm was rated against 14 sets of evaluation criteria (see Fig. .6 overleaf). These 14 sets were:

1. Depth of 'home office' resources and support, including planning, architecture, engineering, soils engineering, landscape architecture and project management; (15)

2. Key personnel to be assigned, their qualifications and experience; (10)

3. Recent experience in programmes of similar type and magnitude (successful management of large-scale projects); (10)

4. Specific experience in geographical location where project is to be implemented, and demonstrated knowledge of costs and conditions in the area; (9)

5. Experience in all elements of infrastructure, including: roads, water and sewerage systems, power and communications, landscaping, soils, etc.; (9)

6. Suitability of methodology; (8)

7. Financial stability; (8)

8. Knowledge of urban planning, urban design and community development; (7)

9. General experience and demonstrated knowledge of customs and conditions in the geographical area in which the project will be implemented; (4)

10. Experience in community support facilities, including health and educational facilities, shops and mosques, etc.; (4)

11. Understanding the project; (4)

12. Capacity of firms to accomplish work in required time; (4)

13. History of working relationship among team and/or joint venture; (4)

14. Extent of conflicting current commitments; (4)

Each set of criteria was in turn given a factor of importance (the number in brackets for each item above), with all 14 adding up to 100. Each firm was rated on a scale of 1–10 for each set of criteria. The scale breakdown was as follows:

1–2 points	poor
3–4 points	fair
5–5 points	good
7–8 points	very good
9–19 points	excellent

During the interview process a written brief or description of the project was given to each firm to allow a basis for the preparation of a presentation.

One of the results of applying the 14 items of evaluation criteria was an indication of which firms were strongest in leadership qualities. Those firms that were graded highly on items 1, 3, 4, 8, 9 and 13 could be viewed as having this capability. This was an especially important consideration,

particularly for the housing project, since the scope was such that it required team effort.

When all firms had been graded, a final graphic matrix (see Fig. 7 overleaf) was developed as a summary of the evaluation process. A graphic tool such as the one shown in Fig. 7 is a good presentation device for those not involved in the evaluation process. It should be realised that the final 'scores' of the rating systems in themselves are not important. They are only an indicator and should serve to reinforce the conclusions that the evaluator had arrived at during the evaluation process. The real value of an evaluation process lies here in that it causes the evaluator to consider the meaning of each separate criterion and how it affects the result. It gaves him a better understanding of the whole process.

Selection strategies

On very large projects, where the scope of work is of such a magnitude that no one single A/E consultant could possibly perform all the required tasks, there is a need for the selecting organisation to develop strategies of selection which will result in an organisation that can perform successfully.

To develop a proper strategy, it is important to realise that firms are merely composed of people and that a firm is only as good as the quality of the people therein. It is the most important resource that a company can bring to a project. It is therefore essential, in the evaluation process, that there is some method of appraising personnel. A firm can be asked to commit certain personnel to the project and this can be done within the RFP (Request for Proposal). but it is important that during the evaluation of firms this potential or lack of it is identified. Any good strategy developed for A/E consultant selection should take this into account.

In the housing project, a number of different strategies were developed and evaluated before a final one was adopted. The development of a strategy for selection can change according to the requirements of time, quality and cost. The main requirement was to select a group which could work as a team to carry the project through from inception to completion. This could be done by either of two forms of grouping, the Team Approach or the Specialist Approach.

The Team Approach. Here the objective is to select a group of firms of which each has an area of expertise. Together, they form a team which has the capabilities of performing the work competently. In this case, during the evaluation process, all firms are combed for a particular capable team leader. The team leader will provide the

	EVALUATION CRITERIA	STANDARD SCORE	FACTOR OF IMPORTANCE	FINAL SCORE POINTS	COMMENTS
1	Recent experience in programmes of similar type and magnitude (successful management of large-scale projects).		10		
2	Specific experience and demonstrated knowledge of costs and conditions in the area.		9		
3	General experience and demonstrated knowledge of customs and conditions.		4		
4	Experience in all of the elements of infrastructure including roads water, and sewer systems, power and communications, landscaping, soils, etc.		9		
5	Experience in the needed facilities.		4		
6	Knowledge of urban planning, urban design and community development.		7		
7	Understanding of project.		4		
8	Key personnel to be assigned, their qualifications and experience.		10		
9	Capacity of firm to accomplish work in required time.		4		
10	Suitability of suggested methodology.		8		
11	Depth of home office resources and support including planning architecture, engineering, soils, landscape architecture and project management.		15		
12	Financial stability.		8		
13	History of working relationship among team and/or joint venture.		4		
14	Extent of conflicting current commitments.		4		
	GRAND TOTAL		**100**		

NAME OF FIRM OR CONSORTIUM:
ADDRESS:
DATE:

NOTE:- Standard Score × Factor = Final Score
- Perfect Score equals 1000
- Evaluation Criteria and factors could be modified to suit each specific project.

Fig.6: CONSULTANT EVALUATION SHEET.

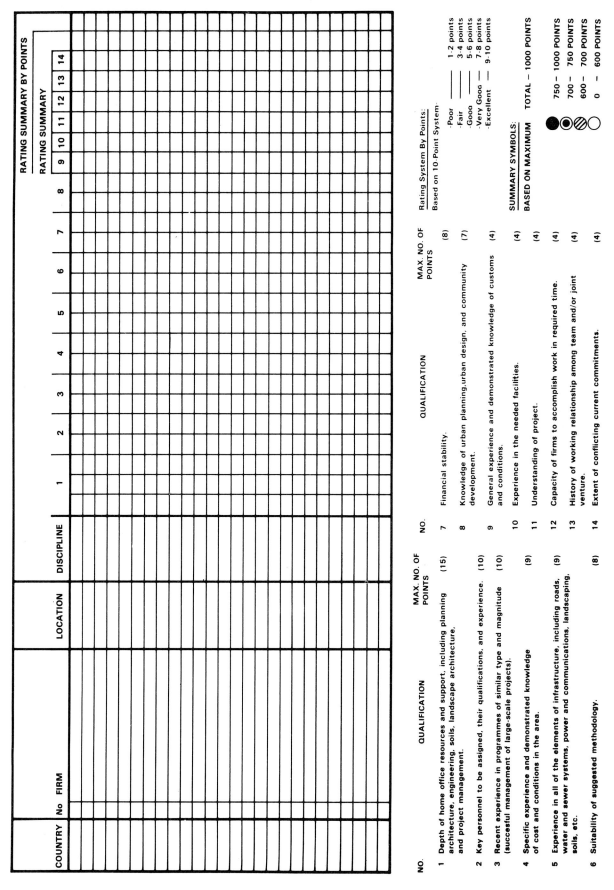

Fig.7: CONSULTANT EVALUATION MATRIX.

'system' or 'process' within which all the other firms will work. The team leader's firm has within it the general project manager and usually performs the project control and programming functions, in addition to carrying out its own specialist tasks.

The Specialist Approach. The objective of the specialist approach, on the other hand, is to select a firm which is very strong or specialised in a certain discipline and then to develop a group of firms which are specialised in complementary areas. One of the specialised firms takes responsibility for the project and uses the input of other firms as it requires. The specialist approach becomes applicable to projects that have as 'core' objectives very specialised requirements. A current example of this type of project is the Space Programme (NASA) in the United States which, because of the specialised nature of technology in many of its programmes sets up groups of this nature. Another example is that of the many US Government military contracts, which, being on the frontiers of research and development in certain areas, must use the firm which has developed a technology as specialist throughout.

6
DESIGN PHASE

This chapter, in defining the design process, makes no attempt to deal comprehensively with the subject. Rather, discussion of it will be in the context of large scale projects and of how the process differs when the client is a public sector government agency.

Design process

In analysing the delivery process of a large scale project, it is helpful to view the process in terms of the major stages which occur from inception to completion. Though there are a multitude of activities occurring throughout the delivery process, there are four major stages (See Fig. 8 overleaf). The first stage involves efforts of analysis that will eventually lead to an understanding of the owner's needs and requirements and of other factors which will be involved in bringing that programme through the delivery process. The second stage is known as the 'translation stage' or, more commonly, as the design phase. It is during this stage that collected and analysed information relating to the owner's needs and requirements is translated and assimilated into a design solution. Stage three is the actual execution of a design solution into its real form, and stage four – 'realisation' – deals with the actual usage, operation and maintenance of the project. In this chapter, the author will deal primarily with phase two, and its interrelationship with the delivery process as a whole. In describing the design process there are as many opinions as there are people as to what is involved. Take, for example, the first step in the design process: programming. Edward Agostini of Becker and Becker Associates, a New York architectural firm, describes the programming process as 'information – not design.' (Agostini 1968). Herbert McLaughlin of Kaplan/McLaughlin/Diaz, architects/planners, describes programming as 'design; particularly contemporary programming which has become increasingly comprehensive and

complex.' (Palmer 1981). There are many different and divergent classifications of the elements of the design process and the relative importance of each, with each classification having a degree of relevance.

The design process can best be described in general terms as a process of 'synthesis' where the programmatic requirements of the client have been clearly defined and translated into an actual building design to be constructed. Within this general context of the design process there are several definable stages: a programming stage, a preliminary design stage, a design development stage and finally a construction document stage. In many parts of the world there could be several 'pre' or 'post' stages tacked onto the process described above. For instance, many firms in North America would include 'contract and tender document preparation' as part of the design process. As described earlier, a base line study would be a preprogramming effort that could be performed by the owner or by a consultant hired by the owner.

It has been the experience of the author, that the major phases of the delivery process change somewhat in the context of large scale projects. These changes occur out of the necessity to fully understand the project at its early stages and to develop systems of control which will govern the development of ensuing stages of the delivery process.

Richard Jacques describes the process thus:

> To achieve the goals of time, quality and cost, the architect must view the design process somewhat differently. Each major parameter is considered during each major phase of the project; only the level of detail related to each concern increases from phase to phase. This approach applies to concerns in planning, design and construction and has as its objectives to: ensure that no major decision is overlooked; allow architect and consultants to get projects under complete control early; permit decisions made in one phase to serve as boundaries or parameters for those to be made in the next phase;

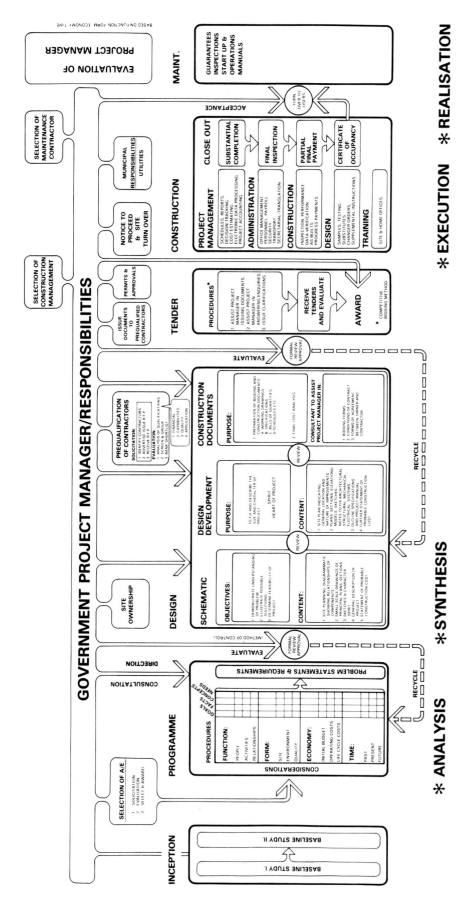

Fig. 8: PROJECT PROCESS.

assure that commitment to a specific solution is made at all points along the way and not reserved until the end; and to avoid the constant redesign when new information is introduced late in the design process. (Jacques 1976).

In the context of the total delivery process of a large scale project, each step of the delivery process is heightened to a greater degree of complexity because of the size of the project. Because large scale projects require large scale inputs of time and resources, a primary consideration is to develop a systemised process that guarantees maximum utilisation. It is this factor, the huge input of time and resources, which causes modification to the conventional delivery process. There is a greater emphasis on the analysis phase, which becomes greatly expanded, taking into consideration more than a design scheme as a result. There are other important factors that must be considered: the systems of control, management, delivery best suited for a particular project; evaluation of different alternatives and the cost/benefit of each. Due to the huge costs and amount of time involved, there must be a clear understanding of the total scope of the project earlier. Before large investments are made in producing a design, there should be an analysis of all viable alternatives so that the one chosen is the best suited.

It seems timely to ask the question, 'In what ways does the systems approach alter the design process?' That can best be answered by looking at the delivery process of a conventional building using a traditional design approach.

With this process, the design phase traditionally encompassed the elements of: (1) programming, (2) preliminary (or schematic) design, (3) design development, (4) construction documents. Under a systems approach the design process is broken into three critical steps: schematic approach, schematic design and design development.

The author here is not advocating a change in the conventional method of design delivery. However, it has been his experience that the schematic phase, if broken into two parts, allows the designer the flexibility to deal with the overall concepts of the project schematic approach step before dealing directly with them in relation to building form (schematic design).

Schematic approach

The 'schematic approach' is the first step into the translation and assimilation stage. The term was developed by the State University Construction Fund (SUCF) of the state of New York and was used to describe the first of two major tasks performed during schematic design: developing a

very generalised approach or concept of the project before proceeding to the (what was normally understood as) schematic design information. SUCF found, through their considerable experience in facility planning, that such a step was necessary to '... focus the owner's and architect's attention on the basic concept of the project and that, by getting collective agreement at this point, the ability to proceed smoothly with further project development would normally be assured' (Jacques 1976).

The schematic approach encompasses three areas of conceptual thinking: the design concept, the site concept and the systems concept (Jacques 1976).

The *design concept* is the assimilation of the programme requirements which relate to the physical spatial design–its general qualities and order. It deals with the issues of form and mass, access and circulation, functional relationships within the project and also, within the approximate region, uses of materials and textures, light, shade, interior design, and relationships of exterior to interior (Jacques 1976). It provides the unifying framework for later more detailed decisions.

The *site concept* is the assimilation of the programme requirements concerning the physical site and its relation to the building design. It deals specifically with the latter, with the relationship of the project to the existing elements of the site and the impact that the project will have on them, patterns of circulation, access and egress for all modes of transportation, orientation of the project elements to the site and to each other, subsurface conditions of the site, utility locations and landscaping.

The *systems concept* is the assimilation of types of systems which relate to the design concept: systems of physical support, enclosure, energy use or conservation, systems of construction, comfort support, life safety and the like.

These three 'concepts' make up the schematic approach. Each should be developed to the point that clearly shows how each will relate to and affect the bulding programme. The object is to keep communication in a conceptual form that allows the flexibility to deal on a general level, yet allows both the architect and owner to comprehend the overall project.

Schematic design

Proceeding to the schematic design phase of the design process means that there is agreement between the architect and the owner on a conceptual solution to the programme. Schematic design further develops the 'conceptual solution', careful to maintain its integrity, into an actual building design. Throughout this translation process, rigorous testing and checking of the actual design against the conceptual solution and the original programme should be done. The schematic design is complete, when every element of the project has been formalised and described. At this stage, the buildings' construction systems should be clearly defined: structural, foundations, enclosure, vertical circulation, environmental control, security, access, egress, site construction, interiors and graphics should be integrated into the overall design.

It is important to realise that with the completion of the schematic design, the architect enters into a period of refinement. The overall context of the building and the systems involved is established and all later decisions are made within this framework. The project is communicated to the owner through drawings and written text that clearly delineate all aspects of the project. The general form and shape of the building is clearly expressed, with all related systems well thought out and integrated.

Cost estimates during schematic design

As, once all work done after the schematic phase has been completed, there will be a refinement process, it is imperative that the owner fully understands all implications of the design. This means that he must be informed of the probable cost of the design, of a schedule for construction, and any problem areas that may need further clarification. Any following steps in the design process should deal only with further refinement and documentation of the design, with minor decision-making occurring within the framework set up by the schematic design.

If the owner has developed a base line study the designer's task in establishing a budget is somewhat simplified. The analysis of costs in the base line study will provide a reasonable basis for establishing a budget in the programming process. In cases where the owner has done no analysis of his needs or requirements, the establishment of a budget becomes a primary task of the programming process. As the design proceeds from the programming phase into the schematic design phase, the conceptual design develops into an actual design. As the design takes on tangible form, it becomes easier to measure quantitatively. The same occurs in developing cost estimates. The tendency is to go from general cost budgets to specific cost estimates. The important thing is to establish a realistic budget early in the design process. As Pena points out,

> Realistic budgets are predictive and comprehensive. They avoid major surprises. They tend to include all the anticipated expenditures as line items in a cost estimate analysis. The architect must look to past

experience and published material to derive predictive parameters (Pena 1977).

During the schematic design process costs are developed around three parameters: (1) a reasonable efficiency ratio of net to gross area, (2) cost per square foot escalated to mid-construction; (3) other expenditures as percentages of building cost. With the exception of site acquisition costs, all major costs are linked to building costs.

Design development

The design development phase of the design process is a further refinement of the schematic design. In this phase, refinement takes place within the framework of the design developed during schematic design. All major decisions have been made during the programming and schematic design phases, with only minor decisions occurring in this phase. Design development can best be described as a 'filling out' stage. If the design development stage has been successful, the architect will find that preparing construction documents will be a process of finalisation and production. Upon completion of design development, the architect should have the following:

- A fully developed site plan showing building location and access, circulation, planting, grading and utilities; all fully thought out and interrelated;
- A fully developed architectural solution, with all spaces firmly located and dimensioned and with all sections and elevation features thought through;
- All building systems (mechanical, structural, electrical, and equipment) fully developed and carefully integrated;
- A set of outline specifications;
- Cost implications of all design decisions considered and recorded;
- A confirmation of tentative decisions regarding tender packages and tender forms (this last criterion becomes necessary if the owner is considering bidding the work in different packages as in a phased design and construct delivery process). (Jacques 1976).

Construction documents

If the design has progressed from stage to stage successfully, then the construction document phase will be devoted mainly to production. Preparation of construction documents is a process of documenting the physical spatial design. Every aspect of the design is translated into a form that can be communicated in a standard language. During this production phase, other important activities will occur. A schedule for tendering and construction should be finalised and agreed to by both the architect and owner. Prequalification of selected contractors should begin.

Because the construction documents describe every aspect of the design in great detail, it is possible to develop an accurate cost estimate at this point. This final estimate serves as an indicator of what the total cost of the project will be. It should be reviewed carefully by both the architect and owner to ensure that all issues or questions relating to cost have been answered. The final estimate should make allowances for the schedule of construction and how it and market conditions can influence costs.

Design management system

A large scale project requires close co-ordination, timely decisions and close monitoring for successful completion. A comprehensive management system should be instituted for the project, which will utilise experienced managers, the latest computerised tools of management to aid them and also a clearly defined line of authority which assigns responsibility so that each team member knows exactly what is expected of him.

This will be achieved by concentrating authority and responsibility in the office of the Project Director. The Project Director will have final responsibility to the owner for the performance of the entire team. Working under the Project Director will be Project Managers from the owner's agency or department and also the Consultants. Supporting these managers will be an experienced team of programmers, cost-estimators and other support personnel.

Management programme

The Management Programme should include the following:

1. *Project manual*
At the beginning of the assignment, concurrent with the preparation of the project inception report, the consultant should prepare a project manual for the team which will include background technical and operational information that will be used to orientate and inform everyone on the team. An important segment of this manual will include the chain of command, channels of communication, methods of communication (e.g., memos, letters, transmitters, reporting procedures). A standard format will be established for the various monthly reports and schedules. The manual will be updated and expanded as the work progresses and the need arises.

2. *Project work plan*
After scrutinising the project site and specifying the detailed site delivery plan, the consultant should prepare the executive work plan for the whole project, specifying all work tasks and a time schedule for the implementation of these tasks, and defining the percentage of these tasks related to the whole scope of the work. This plan will include detailed plans and schedules which project the activities of each of the team members on each of the tasks. This early planning and scheduling format will form the basis for the computer scheduling of the team's complete activities over the duration of the work. After the approval of this plan, it will become the basis of the project progress and should be updated every two months, at least.

3. *Monthly reports*
Each month the management team should summarise its activities in a report to the Project Director. This report should include those items which will form the basis for evaluating progress against the plan. This should include recommended actions where action is required. The report should include the number of man-hours that have been utilised.

This concept of monitoring actual progress against a plan extends to the three principal variables of a construction project: cost, time, and quality. In each case, the team will develop a basic plan, expand it in detail as required as the project proceeds, gather information on that variable each month, report on the status, make recommendations as to where action should be taken, and carry out the resulting decisions.

4. *Invoice reports*
In accordance with an executive work plan approved by the owner, the Consultant should define all work tasks and determine when the percentage of these tasks are completed. The corresponding fees will then be calculated. This information will be submitted in a report along with an invoice. This report should also include all documents proving the completion of these tasks during that period.

5. *Cost control*
The team begins with the formulation of a carefully considered budget, using the budget estimate as a basis. The budget is classified further into a budget for each project stage. Information is then collected through a series of estimates. This estimating information is compared to the planned total loss and variations if applicable and results in recommendations for client/consultant re-evaluation.

6. *Schedule control*
The team should prepare a project master programme which is amplified into design schedules, procurement schedules and potential construction schedules. Each month's progress should be com-

piled, and the overall project programme will be re-run, resulting in forecasted completion dates for each of the major activity areas. If these dates violate the plan given in the master programme, recommendations will be made for changes in priorities, design or re-scheduling.

7. *Quality control*

Quality control refers to design quality and technical quality. In the design quality area, the team should utilise a system of in-house reviews of the design progression. In the technical quality area, the consultant should be requested to establish a check-out team for the project, composed of experienced technologists. This team will initially prepare the technical section of the project manual and then will review each drawing package, specification and contract document, prior to its release, for co-ordination, accuracy and constructability. No drawings should be released until this team of senior personnel is satisfied that the project is of the highest quality capable of being achieved by the team.

8. *Financial control*

In order to keep the Project Director informed as to the project's financial situation, the consultant should be requested to submit invoices and anticipated expenditures, which should not deviate from the financial plan presented in the Project Inception Report.

Project Inception Report

At the end of the project's initiation, the team should present a Project Inception Report to the Project Director which will detail the methodology the team intends to follow to complete the project. The Project Inception Report should include the following items:

Work assignments
Project schedule
Project personnel
Project financial plan
Tender and contract award procedures
Procurement recommendations
Monthly progress report
Quality control procedures
Training programme.

(Ministry of Interior 1980)

7
CONSTRUCTION TENDER

Tender procedure is the most critical part of a project. Project Directors should treat it with caution and care. The Project Director should supervise and monitor its progress, step by step, with some steps personally performed by himself (such as the short listing and final recommendations) and he should not delegate responsibility in these matters to anyone.

Tendering procedures

The tendering stage should include the procedure for pre-qualifying contractors receiving tenders and making contract awards.

A chronological listing of the tendering stage elements should be contained in a Tender Report. The time frame for these elements should be included in the schedule.

Procedures for construction document preparation, final estimates and construction contracts should be included in other documents prepared as part of the design stage,

The Tender Report and the schedule present a synopsis of the tendering stage process, including the objectives of the evaluation and the anticipated results. The solicitation process is a method for identifying contractors for evaluation. The evaluation process is a method for screening identified contractors, analysing their qualifications and short listing them to obtain the optimum number of contractors qualfied to bid for the project.

The evaluation process also includes the recommendations made to the decision maker and an explanation of the later pre-qualification of contractors by the decision maker.

In the tendering process section, Contractor notification and verification, distribution of tender documents and the tender opening should be explained.

The tender award process includes analysis, negotiations and recommendations for bid award.

The tendering stage procedure includes the steps from solicitation through to contract award. (See Fig. 9 overleaf.) The steps are listed below:
1. Solicitation process (contractor identification)
    (a) project criteria analysis
    (b) advertisements
    (c) response to adverts/issue questionnaire
    (d) registration questionnaire effort by contractor
2. Evaluation process (prequalify contractors)
    (a) screening and grouping
    (b) analysis of qualifications
    (c) short listing
    (d) recommendations to decision maker
    (e) contractor selection
3. Tender process (receipt of tenders)
(a) contractor notifications
(b) verification of notification by contractors
(c) distribution of tender documents
(d) pre-bid meeting
(e) response to enquiries
(f) tender submission
(g) tender opening
4. Contract award process (award of contracts)
    (a) analysis and clarification of tenders
    (b) recommendations for tender award
    (c) negotiations
    (d) tender award
    (e) contract award.

The solicitation process identifies contractors to be evaluated and prequalified. The following are project parameters:

1. Building types: housing, schools, health and recreation facilities, mosques, shopping, power plants, sewage treatment plants, infrastructure, highways.
2. Construction conditions: climate, mobilisation, logistics, labour, procurement, national regulations.

Advertisements are expected to generate responses from contractors who receive pre-

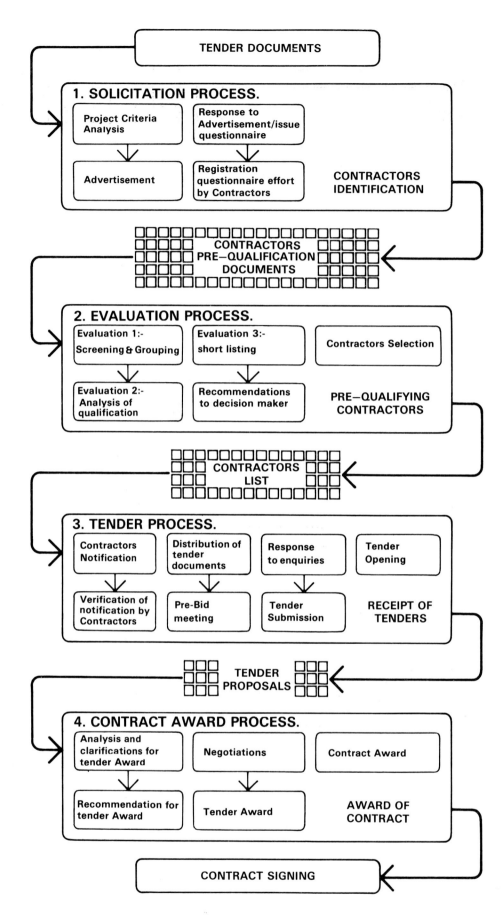

TENDER DOCUMENTS

1. SOLICITATION PROCESS.

Project Criteria Analysis

Response to Advertisement/issue questionnaire

Advertisement

Registration questionnaire effort by Contractors

CONTRACTORS IDENTIFICATION

CONTRACTORS PRE−QUALIFICATION DOCUMENTS

2. EVALUATION PROCESS.

Evaluation 1:- Screening & Grouping

Evaluation 3:- short listing

Contractors Selection

Evaluation 2:- Analysis of qualification

Recommendations to decision maker

PRE−QUALIFYING CONTRACTORS

CONTRACTORS LIST

3. TENDER PROCESS.

Contractors Notification

Distribution of tender documents

Response to enquiries

Tender Opening

Verification of notification by Contractors

Pre-Bid meeting

Tender Submission

RECEIPT OF TENDERS

TENDER PROPOSALS

4. CONTRACT AWARD PROCESS.

Analysis and clarifications for tender Award

Negotiations

Contract Award

Recommendation for tender Award

Tender Award

AWARD OF CONTRACT

CONTRACT SIGNING

Fig.9: TENDER PROCEDURE.

42

qualification questionnaires, which are also referred to as the Contract Registration Forms. The format for this questionnaire is based on organisation, financial resources, physical resources and experience.

When the questionnaires are returned, the information can be placed in a computer system which will log data and analyse the contractor's qualifications. This is the evaluation phase. The three-part evaluation process should be designed to screen and group contractors for projects for which they are qualified to build. Contractors may be eliminated by the following criteria:

1. Classification by concerned agency (Ministry of Public Works)
2. Contractor registration deadline
3. Annual company sales
4. Financial data completed
5. Size of previous guarantees
6. Limitation of sub-contracting to 40% maximum
7. Experience on similar types and sizes of projects.

The analysis of qualifications takes place in evaluation two. The following weighting system is comprised of both qualitative (has practical authenticity) and quantifiable (has objective numerical quantity) information.

Financial Resources	35
Physical Resources	20
Experience	45
Total	100

A short list of ten to twenty contractors could be presented and could be reduced by the decision maker to include eight to twelve contractors in the invitation tender.

An Award Committee will then analyse and recommend tenders for later contract awards. The pre-qualification process effectively evaluates the contractor's ability to fulfil his contract. It also ranks him qualitatively with his peers.

This will allow the committee to analyse the final variable, which is price, and make recommendations for the award of the contract.

Solicitation process

The solicitation process should be designed to identify the appropriate number of qualified contractors to bid on a project. The sequence of steps required to identify a contract or to bid for a project is as follows:

1. *Project criteria*
Establish evaluation guidelines. Establish the project parameters which impart the criteria for evaluation.

2. *Advertisements and response*
Place in a variety of newspapers and trade journals and provide contractors with registration questionnaires.
3. *Receive and log questionnaire*
Input data from the registration form for analysis.

Project criteria

The parameters which determine evaluation criteria should be established from an analysis of the project and the construction environment involved. The main elements are project type, project size, and project location. A contractor's financial status and physical resources, such as management, manpower and equipment, will determine the forces which he can apply to the project.

The construction environment requires the contractor to be familiar with the local conditions such as climate, mobilisation, logistics, labour, procurement, and regulations.

These factors, along with international peculiarities, make working experience under similar conditions the most crucial factor in a project. The parameters from which the evaluation could be established are:

1. Demonstrated experience with projects of similar size
2. Similar working conditions
3. Similar financial demands
4. Similar physical resource requirements

Advertisement and response

The solicitation of interested contractors begins with placing notices in publications such as technical journals and national newspapers and inviting them to register for pre-qualification. The selected publications should provide access, not only to contractors but also sub-contractors, suppliers, manufacturers and transporters who will also be major participants in the project.

The advertisements will generate responses from interested contractors who should be required to pick up their questionnaire form in person. A format should be used for contractor identification, and will be completed as the registration questionnaires are collected. This format should include the following information:

1. Log number
2. Name of contractor
3. Address
4. Telephone number
5. Telex number
6. Contact person
7. Name of person receiving form
8. Signature of person receiving form
9. Date

Registration questionnaire

The contractor registration questionnaire is the document used to evaluate contractors for pre-qualification and short listing to bid on specific projects. It is also provided for identification. Additional information on the questionnaire will be helpful during the bid analysis phase. The questionnaire establishes a consistent contractor profile. The questionnaire could be outlined as follows:

Introduction
Provides the project description and instruction.
Organisation
Identifies and classifies the contractor's organisation.
Financial
Determines the contractor's financial capability and solvency.
Physical Resources
Determines the manpower, physical resources and mobilisation capability.
Experience
Determines the contractor's ability to deliver the project from experience on similar projects and under similar circumstances.

Receive and log questionnaires

1. *Logging questionnaires*
It is useful to log the registration forms as they are received, for identification, filing, and as a check list to see which contractors have been recorded by the computer.
2. *Computer information system*
The computer information system should have the capacity for data input, storage, processing and output in the form of reporting. Data input is made for identification and classification of the contractor. Data is stored in the system for future identification of contractors, processing and reporting during any part of the pre-qualification, tendering and award phases. Data processing is done during evaluation analysis and the short listing evaluation. The data output report which can be generated after this initial input is a contractor list.

Evaluation process

Screening and grouping

Evaluation One should provide a list of contractors who have the qualifications to be analysed for pre-qualification. Also, it should identify group contractors who are capable of construction by projects.

Screening eliminates contractors who have not met the minimum criteria of submitting valid contractor registration data. It also eliminates contractors who do not have the experience, capability and organisational size for the project. Screening criteria for grouping contractors should be based on annual sales turnover, previous guarantees and experience.

The projects to be undertaken by a contractor should not exceed a quarter of his annual turnover; if a contractor has not had guarantees of at least half the project size, there is great concern about his ability to function at the given project level. Similar construction experience is an obvious basis for evaluation, since other criteria may vary in importance for different contractors.

Analysis and qualifications

Evaluation Two provides a list of contractors, ranked in an order suitable for short listing which determines the contractors suitable for the project. The contractor, having passed the screening, and having obtained approval, is a candidate for pre-qualification. Evaluation is then developed by:

Evaluation criteria
A weighting system for criteria
Scoring guide lines
Evaluations by computer information system

Short listing

The objective of the short listing is to provide the optimum number of bidders for the project. Tender documents will be issued only to those who were short listed for the project (or for each part of a multisited project).

Short listing involves selecting the optimum number of prospective tenderers. Contractors can also be selected according to the building system they are capable of constructing, if required. Contractors may also be grouped so that the firms who qualify for the very large projects do not bid against smaller firms that are qualified for the smaller projects. This grouping system (for multisited projects) allows companies to bid against competitors with similar qualifications.

The optimum number of contractors normally asked to bid on a project would be from six to twelve. This should provide a minimum of four submitted bids per project. A much larger bidders' list would discourage participation by many qualified contractors. The selection of the exact number for a short list will depend on the clusters of scores in the list of ranked contractors from Evaluation Two.

The computer processing system will allow the Project Director personally to rank the contractors by total score, by any of the sub-total scores, and by nationality.

Pre-qualification of contractors

The short list of recommended tenderers for the project will be provided for pre-qualification by the decision maker (client). The Project Director should make it clear that his short list is not final but it is only a recommendation to the client. There should also be a list of alternative bidders if possible.

Contract award process

Analysis and clarification of bids

The bid evaluation committee will consider bids financially and technically. The committee may seek assistance from experts during the evaluation process and will also use the conditions of the specifications as guide lines.

The specifications set forth the bid proposal requirements. In short, proposals must be signed by an authorised person representing the contractor's interest and the bid prices must be entered in both words and figures. Corrections of errors will be made but the overall bid price will not be allowed to be revised. All committee results will be kept in writing.

Recommendations for tender awards

At this point the committee will have enough finalised information to make equitable recommendations to the decision maker who will decide on the contract awards.

Contract award

The Owner will make the bid awards and return provisional guarantees to unsuccessful bidders. It should be made clear that the Owner reserves the right to reject any or all the bids and may ask for re-tendering (Ministry of Interior 1981).

8
CONSTRUCTION PHASE

Contract strategies

As the construction process involves the greatest outlay of manpower and resources, and as it is a period when both the owner and the contractor face the greatest amount of potential risk, there is a wide range in choice of construction contracts to accommodate the risks in the most equitable way for both contract partners. However, each contract type has specific requirements and is usually best suited to particular situations or where certain results are required. Generally, construction contracts can be grouped into two main categories. There are competitive bid contracts and owner-contractor negotiated contracts.

Competitive bid contracts

Of competitive bid contracts, the most common is the *lump-sum* contract where a contractor agrees to carry out work for a fixed amount of money. The contractor assumes the risk of completing the work satisfactorily according to the conditions of the contract, regardless of whether the actual cost comes in higher or lower than the agreed upon price. In a lump-sum contract, the contractor assumes the greater risk, but the owner indirectly pays as contractors will estimate a high contingency in their price for unforeseen circumstances. However, this contingency can be affected by market conditions, need for work, and the competitive nature of the bid. A lump-sum contract is also only suitable for a project where the work can be accurately and completely described. Without knowing the total scope of the work described, contractors are reluctant to bid on a lump sum basis.

Another commonly used competitive bid contract is the *unit-price* contract. It is particularly suited for construction projects where the total scope of work is hard to ascertain, e.g., tunnels, underground excavation. This type of contract is formulated on unit prices given for each type of work required as per the conditions of contract. The total cost of the project is not known until the work is complete. The contractor is obliged to perform the work at the unit rates quoted, regardless of the extent of the work, and payment is made only on the work actually done in the field. For the contractor to competitively bid such a contract, there would have to be enough documentation to allow him to assess its overall magnitude. Of particular importance is a set of performance specifications that detail the quality and methods of construction desired.

Negotiated contracts

Negotiated contracts are used where the owner wants to use a particular contractor or where the nature of the project does not allow enough time to produce a full set of detailed drawings, or where the nature of the work will call for many changes during construction. A common type of negotiated contract is the *cost-plus* type contract. With this type of contract, the owner and contractor negotiate a contract based on a 'preliminary' project scope determined by preliminary design drawings and outline specifications, with the understanding that this will be amplified in future as the scope of the project becomes clearer. However, when negotiating a cost-plus contract, there must be clear understanding on three points:

1. A definite and mutually agreeable sub-contract-letting procedure should be arranged. Competitive lump-sum or unit-price sub-contracts are generally to be preferred by both parties when they are feasible. If the nature of the work is such that competitive sub-bids cannot be compiled or are not desirable, a mutually agreeable negotiation procedure will have to be devised.

2. There must be a clearly understood agreement concerning the determination and payment of the contractor's fee. Fees may be determined in many different ways.

3. A common understanding regarding the accounting methods to be followed is essential. Many problems and controversial issues can be avoided by working out in advance the details of record keeping, purchasing and the reimbursement procedures (Ministry of Interior 1981).

For a project where the scope of work is particularly hard to define, one of the best ways for the contractor to determine his fee is with a *cost-plus-percentage-of-cost* contract. This type of contract is especially useful for renovation work or for work of an emergency nature where there is not enough time to go through the normal bidding procedures. The one drawback with this type of contract is that there is no incentive for the contractor to keep his cost down. In fact, it is to the contractor's advantage to prolong the work. However, every type of contract should be made with a reputable contractor of whom the owner has had prior experience and with whose performance he is satisfied, a factor which becomes particularly important when using this type of contract.

Cost-plus-fixed-fee contracts, on the other hand, do provide the incentive for the contractor to complete the work in a timely manner. With this type of contract, the contractor's fee is constant and is not tied to the total cost of the project. It is thus more advantageous for the contractor to finish the work as expeditiously as possible, since any delay will only cost him more money by keeping his manpower and resources tied up in the project.

There are various forms of *incentive* contracts. However, all provide the contractor with an incentive to keep the cost or time duration of a project to a minimum. With these types of contract, a 'target' estimate for time and cost is arrived at and agreed upon by the contracting parties. Because this target estimate becomes a main provision of the contract, a detailed set of construction documents should be made available to the contractor. An incentive 'bonus' is negotiated with the contractor for either early completion or for lower costs. It is to the contractor's advantage to complete the project as soon and as economically as possible in accordance with the contract documents. In terms of cost, the normal practice is for the owner and the contractor to split the difference in cost savings below the target estimate. In terms of time, the contractor would receive, in addition to his normal fee, a bonus for each day of early completion. The contract can also be written to introduce contractor penalties for each day of delay or for a share of any cost over-runs.

Guaranteed maximum contracts were developed in response to the objection that cost-plus contracts never accurately reflect the total project cost (and therefore in some cases, total fees) until after the project is complete. With a guaranteed maximum contract any arrangement of fee basis can be used. However, a guaranteed maximum price for the project is agreed to by the owner and the contractor, prior to signing the contract. With a guaranteed maximum stipulation in the contract, the contractor will have to bear all costs that exceed that amount. Again, this type of contract requires a detailed set of construction documents. To provide additional motivation to the contractor, an incentive bonus can be included in the contract provisions, whereby the contractor will receive a portion of any cost savings below the guaranteed maximum figure.

Phased design/construction or fast-track systems

Phased design and construction is an approach to the project delivery process that saves time for the owner by overlapping the design and construction process. This type of approach has become popular in North America and in some parts of Europe in recent years as a way of combating cost escalation due to inflation. Using a traditional process of delivery, the steps from signing an agreement with the architect to opening tenders for construction could take years, even if everything went smoothly (Foxhall 1972). It is not unusual for bids to come in over the owner's budget and for a project to be recycled through the design process to bring it in line with the budgeted cost. This can make the period between design and construction even greater. In addition, there is the actual time of construction itself. Because 'fluctuations' contracts are not used in Saudi Arabia, the contractor must consider the cost escalation of inflation during the construction period as an unpredictable risk which is reflected in his bid. Sometimes this can be as much as 50% of the budget estimated depending on the duration of the project and the gyrations of the economy. Phased design and construction can alleviate many of the cost problems by redefining the schedule of the project. During the seventies, a time of rapid expansion in the Kingdom's economy (and the subsequent rapid rise in inflation), phased design and construction was advocated by many A/E consultants as a way for the Kingdom to reduce the effects of the inflation spiral.

Phased design and construction will control cost through the manipulation of schedule. By overlapping design and construction, parts of the project will be built before design is complete in other parts. If this occurs, the owner no longer has a clear idea of the costs of the project before he begins construction. In the traditional linear process of delivery, the owner is able to know, within an accurate estimate, the total cost of the project before commencing construction. This allows him the option of either cancelling the project, losing only design fees, or re-defining his budget constraints. In phased design and construction, the emphasis of cost is switched from developing a 'take-off' estimate prior to tendering, to 'conceptual' estimating which is based on 'high end' cost estimating at the schematic phase of design. The owner has no way of knowing actual costs until the project is complete.

Conceptual estimating is a process of cost analysis where cost data is developed at the end of schematic design. The cost data base is developed from contractor's and manufacturer's prices information as well as the owner's, architect's and project manager's cost data. As costs are tabulated, estimates tend to be on the high side. Usually twenty to thirty 'major items' of the project are priced. This becomes the cost indicator of the 'total project cost' for the owner. Thus, costs can be checked against the owner's budget at an early stage during the project. The assumption is that any deviation or subsequent unforeseen cost will average out well under the initial high estimates. As indicated above, 'conceptual estimating' relies on an established and accurate data base of costs. However, in developing countries such as Saudi Arabia, the cost data base necessary to predict cost in such a manner is usually neither developed nor sufficiently comprehensive. Within a developing market, prices can vary greatly and become dependent on accessiblilty of labour and materials and the location of the project. Much of the reason for high inflation during the midseventies was due to congestion at the country's ports. That, and the ambitious building programme of the second five-year plan, caused shortages of many materials, driving up prices.

It has been the experience of the author, that when accurate 'conceptual estimates' cannot be made at the onset of the phased design and construction process, the process loses credence as a viable alternative over the traditional linear process of delivery. The following occurs in this case:

1. The owner is no longer given an accurate estimate on which he can establish a realistic budget.

2. The owner must assume the burden and cost, at no benefit to himself, for the greater management capabilities he will need to implement a phased design/construction process (phased design/construction can present extreme difficulties of co-ordination).

3. The actual cost of the project is open-ended. The real cost will be determined on a piecemeal basis, and the owner has no protection in the case of cost over-run.

4. As most A/E or CM Consultants are aware of the difficulties of controlling cost in developing markets, many are reluctant to assume the responsiblilities of cost over-runs.

5. The owner, due to the nature of the phased design/construction process, becomes locked into the construction process, regardless of its outcome. When cost over-runs become apparent, his only choices are to redefine his budget or to stop construction. If he chooses to stop construction, he has to absorb the much greater cost of both design fees and the construction in place.

Some of the difficulties of owner vulnerability can be overcome by creative contract strategies that spread the cost risks over all major participants (A/E Consultant, General Contractor) the project.

Phased design and construction is inherently the co-ordination of the design and delivery process, in which management does the following:

1. Identifies the building systems that will be needed;

2. Orders those systems in a logical assembly;

3. Relates that assembly to a time schedule for design and delivery (Foxhall 1972).

The process requires close co-ordination between the owner, the project managers and the contractors. This type of process also demands the following:

1. It requires that contractors be experienced and have a good knowledge of lead-time required for procurement of material and labour in the market.

2. It requires a strong project management team that will schedule the work in a way that can ensure performance on time and within the tender price.

3. It requires that the same level of quality be maintained by all contractors being used.

The cost risks involved in getting the project complete in a shorter period of time outweigh the benefits of early occupancy, except in special cases, e.g. national security. What this implies is that for countries that have developing markets, where cost histories are not readily available and where contractors are relatively inexperienced, the owner is exposing himself to cost risks far greater than the cost benefits provided by the phased design/construct delivery process. As opposed to the traditional linear process of delivery, the owner has a greater management task in a more risk-inherent system without any of the cost benefits. In such markets, where expertise is still in the development stage and costs and performance are still unpredictable, the risks inherent in the phased design/construct delivery process make it a non-viable alternative.

Managing the construction process

As a project enters the construction phase, the owner has a number of options in controlling the construction process. Whatever option he chooses, the element of control is implemented by a methodology of management. Although it has been recommended earlier to use the organisational structure shown in Fig. 3 (p. 12), the owner can choose from a number of different organisational structures, each with a particular application, depending on the nature and size of the project. Choice of a particular type of organisation will also depend on the level of involvement the owner wants for himself and his supporting staff. On large scale projects, the manpower and service skills required to monitor and control the construction process are of such a magnitude that the tasks required cannot normally be achieved with the owner's existing organisation. In most cases, where the owner does not have the necessary in-house skills, he would appoint a consultant to execute these tasks on his behalf. The firm could be organised in a number of ways and provide a varying range of services. For instance, the owner could select an A/E consultant that had the capability of providing CM services also; or, the owner could select a separate CM consultant and A/E consultant and his project director would co-ordinate the two.

Whatever configuration the owner chooses or is advised to choose, there will always be an overlap of four principal roles of management that occur during the construction process. These four management functions are performed by the Owner, the Architect/Engineer, the Project Administrator and the Construction Manager. In order not to confuse the issue, two or more of the four management functions may reside in the same firm. For example, the Owner may use personnel on his staff to act as Project Administrator. As stated earlier, the A/E consultant may also have the capability of providing CM services. Regardless of who will provide each kind of management service during construction, it is important to understand their role, the limits of responsibility and the channels of interface that occur.

Depending on the project, the role of project manager can be assumed by any of the team members; or, for the duration of the project, the role can move from member to member depending on requirements. Ultimately, though, it is the Owner, as a determinant of his requirements, who

selects which participant of the project team will assume the role of Project Manager. A clear understanding between the Owner and the A/E firm over the duties and responsibilities of the Project Manager is important when setting up the relationships and responsibilities of the other members of the management team.

Generally, each participant has certain responsibilities that then remain his and are only amplified if his organisation assumes the overall role of project management. These roles are described as follows:

The Owner. The owner is ultimately responsible for the construction process. He is the key member of the management team, in that it is he who selects and organises the management team, based on the definition of his needs and priorities. These tasks cannot be delegated, but must be carried out with his sustained presence during the construction process. Though there are different strategies that will alleviate some of the pressures of day-to-day management, the owner remains the ultimate decision-maker during the construction process, especially on decisions relating to cost.

The Project Administrator. The project administrator, as the owner's agent, acts as an adviser and implementer on his behalf, using his professional experience in areas where the owner's knowledge or experience in construction matters is limited. If the project administrator assumes the role of project manager then he becomes responsible for the overall direction of the project. If he is merely a representative of the owner, then his major task is to ensure that the owner's interest is best served by all members of the management team.

The Architect/Engineer. As the prime element in extracting the owner's needs and requirements and translating them into a physical design, the A/E firm has a primary responsibility to see that the design is properly executed. However, their role changes from one of primary influence during the design phase to one of supervision and adviser to the owner during the construction process (unless they assume the task of overall project management). The A/E firm that developed the design phase is in the best position to assume the role of construction manager. In fact, if this kind of organisation occurs, the A/E firm's project manager can easily assume the role of a CM as the management principles for the two positions are basically the same.

The Construction Manager. Though the construction manager's primary function is to control the overall construction process, it is advisable to bring him into regular consultation as early as the schematic design stage. These are usually known as 'pre-tender' services. If the Project Management Method, which was developed by the author earlier, is used, then the construction management services are provided within one management entity made up from the owner's and the consultant's staff. The CM, being an expert on the construction segment of the delivery process, can provide valuable input in advising the architect and engineer of decisions made during design which will directly affect the construction process. The construction manager can also provide other pre-bid services that can be of value to both the A/E firm involved in design, and the owner. These services, as listed in *Professional Construction Management and Project Administration* by W. Foxhall, would include but not be limited to the following:

1. Review drawings and specifications, architectural and engineering, construction feasibility of various systems and the possible design implications of local availability of materials.

2. Prepare periodic cost evaluations and estimates related to both the overall budget and to the preliminary allocations of budget to the various systems.

3. Recommend for early purchase (by the owner) those specified items of equipment and materials that require a long lead time for procurement and delivery.

4. Advise on the pre-packaging of tender documents for the award of separate construction contracts for the various systems and trades.

5. Consider the type and scope of work represented by each tender package in relation to time required for performance, availability of labour and materials, community relations, and participation in the schedule of both design and construction procedures.

6. As schedule criteria of design and construction emerge, the construction manager may, with the co-operation of the architect/engineer, work some of the design operations into an overall CPM or other network scheduling operation.

7. Check tender packages, drawings and specifications to eliminate overlapping jurisdictions and eliminate possible gaps among the separate contractors.

8. Review all contract documents to be sure that someone is responsible for the general requirements on the site and for temporary facilities to house the management and supervision operation.

9. Conduct pre-tender conferences among contractors, sub-contractors and manufacturers of systems and sub-systems to ensure that all bidders understand the components of the bidding documents and management techniques involved.

Construction supervision

The essential services required from the construction supervision team are:

1. *Contract document review*. The supervision team should carry out the following studies and prepare a detailed report on them:

(a) Check that all copies of the contractor's drawings are in accordance with the signed construction contract drawings, and that amendments, addenda and correspondence agreed to at the contract stage are incorporated in the construction contract.

(b) Check the structural and engineering dimension and co-ordination between the drawings and other bid documents.

(c) Check the structural calculations and the reinforcement drawings for compliance with the construction contract specifications.

(d) Check the calculations for adequacy and conformity with the contract specifications agreed at the signing of the contract. These checks shall identify all faults and errors which would impair the safety of the structures and the installations.

(e) Check the unit and total prices of the Bills of Quantities for each individual part and for the whole project. Check that these prices are properly distributed among the various items, keep the Project Director informed regarding any anomalies that may occur during the performance of the works and submit appropriate recommendations regarding their technical and economic acceptability.

2. *Planning and scheduling.*

(a) Review the contractor's proposed job site management plan, quality control procedures, and manning and mobilisation plan.

(b) Establish control measures to maintain orderly co-ordination and systematic progress in accordance with approved construction schedule.

(c) Review the contractor's material order schedule and delivery dates to ensure that they are consistent with the approved construction schedule.

(d) Review the contractor's construction plan, its adequacy for meeting contract performance requirements, and its location and layout.

(e) Institute a programme of early surveillance of the contractor's proposed mobilisation plan and operating procedures and its performance and progress, in order to avoid problems and delays in the start-up of construction activities.

(f) Study and evaluate the programme of the works (detailed work schedule) submitted by the contractor both in detail and in its overall compliance with the contract period agreed with the owner. This programme should be prepared according to the critical path method (CPM).

(g) Study and evaluate the contractor's manpower and equipment schedules in relation to the programme.

(h) Use the project schedule as a reference to monitor the progress of the contractors. Estimated quantities and man-hours should be correlated with activities in the schedule to provide the basis for a time phase performance evaluation of installation rates and associated costs. This evaluation should be performed utilising data provided directly from the contractor in conjunction with supplemental quantity surveys performed by the consultant.

(i) Use the master project schedule as the basis for status evaluation and critical path analysis. Progress updates and logic modifications should be made in an on-going manner utilising the interactive capabilities of a computer system. Each month a set of scheduling analysis reports and summaries should be prepared as part of the project progress report.

(j) Install a computer control system in order to maintain schedule, cost and quality data on a periodic basis. Accounting information should be tracked monthly and correlated to percentage information. Exception reports should be generated by the system when: committed costs exceed estimates; actual costs exceed committed costs; actual quantities exceed estimated quantities; and actual unit rates (i.e. man-hours per quantity) or actual wage rates exceed those estimated. Scheduling exception reports should highlight critical path action items as well as trend summaries for the contract.

3. *Soil investigations.* The supervision team should verify the location of the trial borings for foundations and the accuracy of the design in accordance with the relevant drawings. The team should review the soil investigation study, subsequent reports, the bearing capacity of the sub-soil. The team should study the reports prepared by the contractor if difficulties are encountered during construction of the foundations and prepare a recommended report.

4. *Site layout.* The team should check the contractor's planning and layout of the site with regard to surveying, elevation and location of buildings, roads and utilities.

5. *Field work supervision.* The site team should inspect all field work performed by the contractor, should satisfy itself regarding the correctness and conformity of work done with drawings and specifications and should also ensure a good standard of workmanship. Inspection activities should include, but are not limited to, the following:

(a) Checking of lines, grades and bench marks established by the contractor

(b) Excavation

(c) Backfill, soils graduation, compactions, etc.

(d) Utility installation

(e) Concrete emplacement including formwork, surface preparation, cleanliness, embedments and reinforcing steel

(f) Concrete batching techniques

(g) Concrete placement controls including rate of placement, method of placement, vibration and construction joints

(h) Concrete testing, slump, air temperature

(i) Concrete placement, finishing, curing and protection

(j) Structural steel erection

(k) System cleaning, flushing and hydrostatic testing

(l) Installation of mechanical, electrical and instrumentation components

(m) Inspection of field run piping, conduct and instrumentation tubing

(n) Installation of electrical power and control cables including cable tray and duct banks

Ths supervision teams should ensure that materials and equipment are used according to the specifications and good standards of practice.

6. *Testing of materials.* The team should request the contractor to do all necessary tests and submit the results and certificates to ensure that all materials conform to the project specifications and are suitable for the application.

7. *Meetings.* Project management meetings should be scheduled and held on a regular periodic basis with the objectives of creating and maintaining open communications, resolving problems of co-ordination and scheduling future work to attain project objectives.

The supervision team should arrange for construction meetings to be held, for example, at two-weekly intervals, and attended by the contractor's representative. The exact dates for the site meetings should be prepared one month in advance and be submitted to the Project Director on a monthly basis. These meetings will deal with:

(a) The project work completed during the time period preceding the meeting and its relationship to the contractor's programme and the targets set at the previous meeting

(b) The project work to be completed during the subsequent time period and its relationship to the contractor's programme

(c) Delivery situation of major items of materials, plant and equipment

(d) List of drawings, specifications and samples approved since the previous meetings

(e) List of drawings, specifications and samples to be submitted for approval in the subsequent time period

(f) Number of contractor's men on site categorised by trade, etc.

(g) Possible complaints, accidents and problems concerning the project

8. *Status reports.* Schedule and quality data supplied from the contractors should be tracked, verified and reported monthly. The results of this data should be included in a monthly Project Report. Items covered in this report include, but are not to be limited to:

Project status

Significant construction activities; materials and equipment delivery, changes and completion of systems.

Contractors

A list of contractors and their activities for the reporting period, their manpower utilisation, productivity, needs and availability, and material/equipment status.

Schedule

Estimate of percentage of work completed based on the project schedule, current position and future trends regarding milestone events.

Financial

Cost and accounting information including cost to date, committed cost, project cost to complete and estimated cost at completion. Cash flow requirements should be included.

Administration

Project meetings, weather conditions, delays and their causes, visitors, safety and security.

9. *Progress payments.* The team should check and verify the progress accounts for work implemented and materials on site submitted by the contractor. When the team has satisfied itself that the claim is correct and complies with the conditions of contract, and has verified the accuracy of measurements and quantities on the site, a payment certificate should be issued. Payment certificates accompanied by a copy of the contractor's claim should be submitted to the Project Director.

10. *Change orders.* Any claims for additional compensation made by the construction contractors should be reviewed against the construction contract documents to determine the validity of the claim. Should the claim be found to be valid and constitute extra work, an agreement should be negotiated with the contractor, and a contract change order should be prepared for approval by the Project Director. Any claims against a contractor for failure to perform should be negotiated and settled prior to final completion of the construction contract involved.

11. *Shop drawings.* The team should review the specifications and call for all drawings required in addition to the original contract drawings and check and approve where satisfactory such drawings provided by the contractor.

12. *As built drawings.* The team should programme, co-ordinate, check and approve the as-built drawings prepared by the contractor. The team reports and makes recommendations on those aspects of the contractor's documentation which are for decision by the Project Director.

13. *Material substitution.* In case the Owner should require a change or substitution of any one of the construction materials, or of an item of equipment or a service, for which there is no price or description of quality in the construction contract or in the specifications or the bill of quantities, then the Owner should be advised regarding an adequate price and, if so required, an indication should be given of the time factor involved. No substitutions or changes should be made without the approval of the Project Director.

14. *Project records.* Complete project records should be maintained to include the following:

(a) Files for correspondence, reports of job conferences, shop drawings and sample submissions, reproductions of original contract documents including all addenda, change orders, field orders, additional drawings issued subsequent to execution of the contract, design engineers' clarifications and interpretations, progress reports and other project-related documents.

(b) Diary or log book, recording hours on job site, weather conditions, data related to questions of extras or deductions, list of visiting officials, daily activity decisions, general observations and more detailed observations, as in the case of observing test procedures.

(c) Daily records of work accomplished in conformance with contract documents.

(d) Record set of prints of contract drawings, showing all 'as constructed' conditions. These drawings should be given to the contractor at the end of the project so that he may prepare the 'as-built' drawings.

(e) Names, addresses and telephone numbers of all contractors, sub-contractors and major suppliers of equipment and materials.

15. *Lack of contractor co-operation.* If the contractor ignores the supervision team's instructions to rectify errors discovered during the execution of the works, the Project Director should be informed accordingly without delay.

16. *Contractor complaints.* All applications and complaints made by the contractors, pertaining to technical as well as other matters, should be checked and submitted to the Project Director along with comments after they have been studied thoroughly from the technical and financial points of view.

17. *Contract completion.* The supervision team after completion of the project proceeds with the provisional hand-over of the project and, after the expiration of the term provided for in the contract of the contractor (normally 12 months), with the final hand-over.

The supervision team should prepare a report containing a schedule of defects that have become evident in the period of time between the two hand-overs. The report should include details of the means to correct these defects, as well as the corresponding costs and time, and must be presented not later than one month before the date of the final hand-over.

The team must check and approve the contractor's final account and submit a report together with recommendations, which will be submitted to the Project Director, after it has been signed by the Contractor. The final certificate should include remarks concerning the quality of the executed work and all those deductions recommended to be made, as a result of default, violation or bad workmanship that may be discovered in the execution of work (Ministry of Interior 1981).

9
MAINTENANCE AND OPERATION

Approach

The Project Manager should plan, organise and control the development of the management plan to direct, maintain and operate the project, as well as the technical studies required for the scope of work, including the following steps:

1. Step One. The identification of the tasks that must be performed. These tasks include the technical operation and maintenance of the facilities, the provision of the services, and the overall management of the project.

2. Step Two. An analysis of how the tasks can be performed, and of the organisation concepts, location and regulatory issues, schedule and financial implications, together with a plan selected for implementation. The management plan for the selected alternative will include the scope of work for the executing entities or agencies.

3. Step Three. The selection of entities or agencies to carry out the work.

4. Step Four. Implementation and monitoring.

Scope of work

Task identification

The objective of this step is to identify the tasks necessary to provide the technical operation and maintenance of the facilities/non-technical services, and overall project management. In essence, step one defines what must be done.

1. *Select Project Manager*
The Project Director should select a qualified Project Manager in the central office, directly responsible to himself, who will lead the development and implementation of the various steps.

2. *Task definition*
A team effort that could include all consultant staff. The team research and detail the tasks. The Project Director reviews and approves. Tasks should be grouped together.

(a) Technical O & M – overall objectives to be stated. General O & M – task frequency and description should be stated in terms of performance required. Final details for installed equipment will depend on what is furnished and should be included in O & M manuals to be provided by the contractor.

(b) Services – overall objectives to be stated. Tasks to be developed grouped by functions such as, religion (mosques), commercial leasing, police services, waste disposal, fire protection, telecommunications and recreation.

(c) Project management – objectives of these tasks to be stated. Tasks to be developed by groups such as financial, administrative, services, staff training, technical, procurement/inventory control, product performance/evaluation, phasing and the user's orientation.

3. *Interactive work session*
This is a team task with the consultant staff and specialists who will carry out an intensive work session to review and complete the task definitions. This session will cover other tasks.

4. *Task documentation*
A detailed description of each task should be prepared, indicating approximate level of effort required and scope. The Project Director should review and approve the submitted documentation.

Task management analysis

The objective is to analyse how the tasks identified in Step One can best be performed; to develop alternative management/organisational concepts; to develop approaches to regulatory constraints; and to develop the financial tools and analysis necessary to evaluate management concepts. In essence, Step Two defines *who* will do what, *when*, and for *how much*.

1. *Develop alternative management concepts*

The team will develop concepts in conjunction with the Project Director and participate in reviews and refinement with him. This will be a continuation of the interactive work session in the previous task.

2. *Develop approach to regulatory constraints*

The team will investigate and develop an approach in consultation with the Project Director and other officials as needed.

3. *Develop financial and technical evaluation tools*

The team should develop and refine, in consultation with the Project Director, the evaluation tools.

4. *Select and refine management concept*

The team should select and refine the concept to be recommended for approval.

5. *Transition plan*

The team should develop a plan for the transition from the contract warranty period to full user's agency management.

6. *Recommended plan for implementation*

As a summary activity the team should recommend a plan for implementation including project organisation and responsibility, definition (scope of work) for contractors, public agencies, users and other.

7. *Report*

The Project Director should present all findings, recommendations and conclusions in a report to the top management committee of the project.

Selection, assignment and contracting

This is to be carried out by the team with appropriate public and private agencies or entities and with the assistance of the team for implementation of the management programme, based on the conclusions reached in Step Two.

(a) The user's agency will prepare and execute agreements with appropriate supporting agencies and entities.

(b) The user will prepare tender documents for consultants, contractors and concessionaires.

(c) The user will select and contract with the consultants/managers as required.

(d) The user will select and contract with O & M contractors as required.

(e) The user will recruit and install additional management/operational staff as required.

Implementation and monitoring

The organisation established in Step Three will implement the management programme, under the supervision, monitoring and quality control of the user's agency. Periodic adjustments will be required as contracts executed in Step Three expire and as changes in Public Agency Regulations require (Ministry of Interior 1983).

GLOSSARY OF DEFINITIONS

This section includes some of the important terms used in the project management profession. The definitions have been developed from the text of this book and other books listed in the bibliography, especially:

Adrian, James J., *Building Construction Handbook*, Reston Publishing Co. Inc., Reston, Virginia, 1983.

Dell Isola, Alphonse, *Value Engineering in the Construction Industry*, New York, 1982.

Ministry of Interior, *Project Management*, Unpublished Report, Saudi Arabia, 1981.

Analysis	Separation or breaking up of a whole into its fundamental elements or component parts.
Bar chart	A series of bars plotted against a calendar scale. Each bar represents the beginning, duration, and end in time to some segment of the total job to be done and together the bars make up the project schedule
Base Line Study I	A document which includes the requirements of the user; also, it gives a general idea about the project in terms of scope, time, quality, cost and impacts on a country's overall development.
Base Line Study II	This document concerns the project implementation plan and how to make the project a reality. It selects the management approach, the delivery method and the control method.
Concept	Something conceived in the mind: idea, notion.
Construction administration manual	A document which explains the method to be used to administer the construction contracts, sets forth the report and submission formats called for in the construction contracts, and describes the functions to be performed by the project personnel at the various site locations.
Construction documents	A process of documenting the physical spatial design. Every aspect of the design is translated into a form that can be communicated in a standard language.
Construction management (CM)	A system made up of interrelated tasks to co-ordinate and communicate the entire project phases, including design,
	tendering, and project execution, with the objective of minimising the project time and cost, and maintaining the project quality.
Cost	The price paid to acquire or accomplish something. In the context of the base line study, cost is associated with the potential and limitations of the problem. In respect of potential, the amount of resources available for the project. In respect to limitations, the extent of resources available. Cost can become an expression of time, for example, as in financial year budget allocations.
Critical Path Method (CPM)	A scheduling technique based on networking (or logical intertwining of the schedule activities). Once the schedule network has been established, the Critical Path Method traces through the network to determine the criticality of each chain (or logic path) through the network. (A logic path is a combination of one or more activities into a series or chain of events). The critical path is the longest path through the network and thus determines the overall duration of the schedule.
Current status chart	A graphic representation of the current status report.
Current status report	Includes the activity number, activity description, work days remaining to completion, current start or finish of the activity, percentage of completed work days, days ahead or behind schedule and expected start or finish based on progress.
Data	Factual material used as a basis for reasoning, discussion or decision.
Design-build construction process	A single firm provides the project owner with project design and construction services.
Design concept	Ideas intended as a physical solution to the client's architectural problem.
Design development	The design development phase of the design process is a further refinement of the schematic design. In this phase,

	refinement is done within the framework of the design developed during schematic design. Design development can best be described as a 'filling out' stage.
Electronic data processing plan (EDPP)	Explains the system design and the use of equipment found in projects which have computing facilities.
Fact	Information presented as having objective reality; truth.
Forecast	The projection of what will take place by a certain point in time.
Functional Manager (FM)	Provides direction regarding who will perform the tasks, how the technical work will be accomplished and how much money is required to perform the work.
General contracting (GC)	A process by which an owner selects a design firm to develop the construction documents and then engages a general contractor for the total execution of the project who will sub-contract various parts of the project with himself acting as the construction manager.
Goal	The result towards which effort is directed. In the context of a base line study, an objective to be achieved. In the base line process there are two types of goals: *Ideal goals* – a set of goals which if achieved would result in the most ideal solution to the problem. *Real goals* – goals that can be realistically achieved which are set through the analysis process of the base line study.
Limitation	The final or furthest bound or point as to the extent or continuance. In the conext of a problem statement, the limits implied in the problem statement; when tested by analysis, these define the parameters of the solution to the problem.
Logic network	A detailed picture of all the various steps required to complete a task or project. It is a bar chart using smaller more specific task bars and including the interpendencies.
Management	The co-ordination and integration of all resources (both human and technical) to accomplish specific results.
Milestone scheduling	A format indicating individual events or milestones involved in completion of a project. It provides a basic time frame to direct the later more detailed scheduling efforts.
Need	Requirement; something necessary; an indispensable or essential thing or quality.
Objectives	Goals, targets or results to be accomplished by a certain time.
Organisation	The orderly arrangement of people's responsibilities and duties to accomplish the project objectives.
Owner	The owner is the overall organisation represented by the top official, which includes several user's departments. A project will have one owner, but could

	have more than one user. Example: Internal Security Forces Housing Project owned by the Ministry of Interior, represented by the Minister of Interior. The users are the several departments within the Ministry, such as the Police and Border Guards.
Personnel procedures manual	Describes administrative details for handling all personnel matters such as recruiting, travel and visa arrangements, working hours, leave provisions, medical insurance and general orientation.
Plans	The detailed steps to be taken in order to accomplish the objectives.
Policy	The guides and ground rules to regulate project actions.
Position description	A written statement outlining the contents and essential requirements of a job.
Position specification	Identifies the qualifications which are needed by a person filling a position.
Potential	The capability of coming into actuality or realisation. In the context of a problem statement: the potential implied in the problem statement, that when tested by analysis, defines the parameters of the problem's solution.
Procedures	The detailed instructions for carrying out a policy.
Programming concepts	Ideas intended mainly as functional and organisational solutions to the client's abstract ideas, generalised for particular instances.
Project	An undertaking which is to be executed within a specified time, cost and quality.
Project budget	The approved cost estimates for the various aspects of the project, including both direct and indirect costs and contingency. The major portion of the estimate is the facility capital cost.
Project delivery process	The methodology used, based on prior analysis of the problem, in determining the most satisfactory combination of time and resources required for a project delivery system.
Project delivery system	A predetermined set of steps taken, leading to the occupancy and operation of a project that reflects a successful solution of the goals and objectives of the client.
Project funding plan	A plan formulated to provide funds when needed. It is based on both a cash flow analysis of the project budget and on the integrated project schedule, in order to identify expenditure levels each month and thereby establish funding requirements. The project funding flow would indicate the working or invested capital required to be provided each month or quarter in support of the project.
Project management (PM)	An application of a system made up of interrelated tasks performed by various resources, with well-defined objectives

	in terms of scope, cost, schedule and quality.
Project management manual	Outlines the mutually agreed upon procedures governing the conduct of the technical and administrative aspects of the project.
Project Manager (PM)	Provides direction regarding clarification of the project tasks, when they should start and finish to meet overall project goals, and how much money is available to perform the work.
Project organisation plan	Develops a strategy for organising personnel to carry out the project objectives, and delineates the levels of management, various project functions and personnel positions responsible for each function.
Project plan	Identifies the objectives, tasks, duration of tasks, responsibilities and information required to design and construct a project.
Project planning control	A system to serve project personnel by providing a practical comprehensive approach to project administration. The ultimate objective of a project planning and control system is to facilitate management action. In the planning phase, the objectives, policies, plans and procedures are organised to provide a framework for project implementation. In the control phase detailed monitoring permits immediate identification of problems so that the appropriate action can be taken and the desired results achieved.
Quality	Fineness or grade of excellence. In the context of the base line study, quality is an objective or goal to be achieved that becomes a parameter for the problem solution after the analysis process of the base line study has occurred.
Requirement	Something wanted or needed.
Resources	The collective means or materials that can be brought to bear on any particular event (or project).
Schedule	A series of things to be done within a given period. In the context of the project delivery process, there are two types of schedule: concurrent and conventional. *Concurrent scheduling* (also known as 'fast track' or 'phased design and construct') is a scheduling process where different events occur together. *Conventional scheduling* is a linear progression of events.
Schedule analysis report	A written document which is compiled from the computer output and project contracts. Highlighted in this report are the current status of the project, present and future areas of concern, cause and effect, possible solutions to critical areas, and a general overview of the project.
Schematic approach	The first step into the translation and assimilation stage of a project. The first

	of two major tasks performed during schematic design: developing a very generalised approach or concept of the project before proceeding to the schematic design information.
Schematic design	Develops the 'conceptual solutions', carefully maintaining their integrity, into an actual building design.
Scope	The area or extent covered by something. The sum total of all factors involved.
Site concept	The assimilation of the programme requirements which relate to the physical site and its relation to the building design.
Staffing plan	Shows the required positions and working terms for personnel to be located at the project offices. It is, however, merely a guideline for the distribution of manpower through to completion of the project and may be modified in response to operational experience and staff availability or in light of changes in project objectives.
Standards	The defined levels of performance by which accomplishment will be measured.
System concept	The assimilation of types of systems which relate to the design concept systems of physical support, enclosure, energy use or conservation, systems of construction, comfort support, life safety, etc.
Systems approach	The process of implementing the solution to a specific problem within a predetermined system that has resulted from an analysis of the problem.
Synthesis	Composition or combination of parts or elements so as to form a coherent whole. The process of translation or assimilation of the results of the problem analysis in the design process.
Time	The duration of all existence; measured in past, present and future and dealt with in the context of the influences of history, the inevitability of changes for the present and projections into the future.
Turnkey contracting	Where a firm takes responsibility for the construction phase (both time and cost), but in fact sub-contracts all of the work. The sub-contractors have contracts with the turnkey contractor.
User	The department which represents a group of people who will use the project after its completion.
Value engineering	An organised team study of functions to generate alternatives which will satisfy the user's needs at the lowest life cycle cost.
Wants	Things lacking and desired or wished for.
Work breakdown structure	Consists of two basic tasks: 1. Determining the component parts of the project systems and facilities; 2. Determining

the services to be provided or tasks to be performed.

Working schedule report   Indicates when an activity is to start and finish, also when it actually did start

(WSR) and finish. This report is issued at the working level, where the detail of a current status report is felt not to be needed.

REFERENCES

ADRIAN, JAMES J., *Building Construction Handbook*, Renton Publishing Co. Inc., Reston, Virginia, 1983.

ADRIAN, JAMES J., *Business Practices for Construction Management*, Elsevier North Holland Inc., New York, 1979.

ADRIAN, JAMES J., *C.M.: The Construction Management Process*, Reston Publishing Co. Inc., Reston, Virginia, 1981.

ADRIAN, JAMES J., *Construction Estimating*, Reston Publishing Co. Inc., Reston, Virginia, 1982.

ADRIAN, JAMES J., *Quantitative Methods in Construction Management*, Construction Systems Publishing Co., Peona, Illinois, USA, 1981.

AGOSTINI, EDWARD J., 'Programming: Demanding Speciality in a Complex World', *Architectural Record*, September, 1968.

AHUJA, H. N., *Construction Performance Control by Networks*, John Wiley & Sons Inc., New York, 1976.

AL-SEDAIRY, SALMAN T., 'Managing Recommendations for Site Supervision and Management', unpublished report, May, 1981.

ARMSTRONG, BRIAN, *Programming of Building Contracts*, Northwood Books, London, 1981.

BENNETT, PHILIP H. P., *Architectural Practice and Procedure*, Batsford Academic and Educational Ltd., London, 1981.

BREARLEY, A., *The Management of Drawing and Design*, Gower Press Ltd., Epping, Essex, 1975.

BURGESS, R. A., *Construction Projects, their Financial Policy and Control*, Construction Press, London, 1982.

CALVERT, R. E., *Introduction to Building Management*, Fourth Edition, Butterworth Scientific Co. Ltd., London, 1981.

CHARTERED INSTITUTE OF BUILDING, *Programmes in Construction – A Guide to Good Practice*, Chartered Institute of Building, Ascot, Berks., 1983.

CLASS, ROBERT ALLEN and KOEHLER, ROBERT E. (eds.), *Current Techniques in Architectural Practice*, American Institute of Architects, Washington, and Architectural Record Books, New York, 1976

CLAUDILL, WILLIAM WAYNE, *Architecture by Team*, Reinhold Co., New York, 1971.

CLIFTON, R. H., *Principles of Planned Maintenance*, Edward Arnold (Publishers) Ltd., London, 1982.

COX, VICTOR L., *International Construction: Marketing, Planning and Execution*, Construction Press, London, 1982.

COX, WELD, *Managing Architectural and Engineering Practice*, John Wiley & Sons, New York, 1980.

CUSHMAN, ROBERT F., STORER, ALAN B., SNEED, WILLIAM R. and PALMER, WILLIAM J., *The McGraw-Hill Construction Management Form Book*, McGraw-Hill Inc., New York, 1983.

DELL' ISOLA, ALPHONSE, *Value Engineering in the Construction Industry*, Van Nostrand Reinhold Co. Inc., New York, 1982.

DRUCKER, PETER F., *Management: Tasks, Responsibilities and Practices*, Harper & Row, New York, 1974.

DRUCKER, PETER F., *The Effective Executive*, Harper & Row, New York, 1967.

FOXHALL, WILLIAM B., *Professional Construction Management and Project Administration*, Architectural Record, New York, and the American Institute of Architects, Washington D.C., 1972.

FREIN, JOSEPH P., (ed.), *Handbook of Construction Management and Organisation*, Van Nostrand Reinhold Co. Inc., New York, 1980.

HARDING, BONNY J., (ed.), *Management of Large Capital Projects*, Instution of Civil Engineers, London, 1978.

HARRIS, FRANK, and McCAFFER, RONALD, *Modern Construction Management*, Second Edition, Granada Publishing, London.

HART, HAROLD D., *Building Contracts for Design and Construction*, John Wiley & Sons Inc., New York, 1976.

HAVILAND, DAVID, *Managing Architectural Projects: the Effective Project Manager*, American Institute of Architects, 1981.

HAVILAND, DAVID, *Managing Architectural Projects: the Process*, American Institute of Architects, 1981.

HOLLINS, R. J., *Production & Planning Applied to Building*, George Godwin Ltd., London, 1971.

JACQUES, RICHARD G., *Current Techniques in Architectural Practice*, American Institute of Architects, Washington D.C., and Architectural Record, New York, 1976.

KELLY, ALBERT J., and MORRIS, PETER W. G., *Strategies for Managing Very Large Projects*, Arthur D. Little, Conference on Project Management, 1981.

KHARBANDA, O. P., STALLWORTHY, E. A., and WILLIAMS, L. F. (eds.), *Project Cost Control in Action*, Prentice-Hall Inc., New Jersey, 1981.

KLIMENT, STEPHEN A., *Creative Communication for a Successful Design Practice*, Whitney Library of Design, an imprint of Watson-Guptill Publications, New York, 1977.

LEE, REGINALD, *Building Maintenance Management*, Granada Publishing Ltd., St Albans, UK, 1981.

LEECH, D. J., *Management of Engineering Design*, John Wiley & Sons Ltd., London, 1972.

LESTER, ALBERT, *Project Planning and Control*, Butterworth Scientific Co. Ltd., London, 1982.

LOCK, DENNIS, and FARROW, NIGEL. (eds.), *The Gower Handbook of Management*, Gower Publishing Co. Ltd., Aldershot, UK, 1983.

LOCKYER, K. G., *An Introduction to Critical Path Analysis*, Pitman Publishing Inc., Marshfield, Massachusetts, 1977.

MARSH, P. D. V., *Contract Negotiation Handbook*, Gower Press Ltd., Epping, UK, 1974.

MASON, JOSEPH G., *How to Build Your Management Skills*, McGraw-Hill Book Co., New York, 1971.

METZGER, PHILIP W., *Managing a Programming Project*, Prentice-Hall Inc., New Jersey, 1981.

MILLS, EDWARD D. (ed.), *Planning: Architects' Technical Reference Data*, A Building and Contract Journals Book, Ninth Edition, Newnes-Butterworths, London, 1978.

MINISTRY OF INTERIOR, 'A Course in Project Management', unpublished report, Saudi Arabia, 1981.

MINISTRY OF INTERIOR, 'Construction Supervision', unpublished report, Saudi Arabia, 1981.

MINISTRY OF INTERIOR, 'Maintenance and Operation', unpublished report, Saudi Arabia, 1981.

MINISTRY OF INTERIOR, 'Project Inception Report', unpublished report, Saudi Arabia, 1980.

MINISTRY OF INTERIOR, 'Site Supervision', unpublished report, Saudi Arabia, 1982.

MINISTRY OF INTERIOR, 'Tender Procedure', unpublished report, Saudi Arabia, 1981.

MORRIS, PETER W. G., 'An Organisation Analysis of Project Management in the Building Industry', *Build International Magazine*, Applied Science Publishers Ltd., U.K., 1973.

OXLEY, R., and POSKITT, J., *Management Techniques Applied to the Construction Industry*, Granada Publishing Ltd., St Albans, UK, 1980.

PAICE, P. A., *Critical Path Analysis: Basic Techniques*, Longman Group Ltd., New York, 1982.

PALMER, MICKEY A., *The Architect's Guide to Facility Programming*, The American Institute of Architects, Washington D.C., and Architectural Record Books, New York, 1981.

PASCALE, RICHARD TANNER and ATHOS, ANTHONY G., *The Art of Japanese Management*, Penguin Books Ltd., Harmondsworth, UK, 1983.

PENA, WILLIAM., *Problem Seeking: An Architectural Programming Primer*, C.B.I. Publishing Co., 1977.

REYNOLDS, HELEN and TRAMEL, MARY E., *Executive Time Management*, Gower Publishing Co. Ltd., Aldershot, UK, 1979.

R.I.B.A. (Royal Institute of British Architects), *Guide to Group Practice and Consortia*, R.I.B.A. Publications Ltd., London, 1976.

R.I.B.A. (Royal Institute of British Architects), *Handbook of Architectural Practice and Management*, Fourth Revised Edition, R.I.B.A. Publications Ltd., London, 1980.

TOPALIAN, ALAN, *The Management of a Design Project*, Associated Business Press, London, 1980.

WILLIS, ARTHUR J., and GEORGE, W. N. B., in collaboration with WILLIS, CHRISTOPHER J., and SOHER, H. P., *The Architect in Practice*, Granada Publishing Ltd., St Albans, UK, 1981.

WILSON, R. M. S., *Cost Control Handbook*, Gower Press Ltd, Aldershot, UK, 1983.

WOODWARD, J. F., *Quantitative Methods in Construction Management and Design*, The Macmillan Press Ltd., London, 1975.

APPENDICES

The purpose of the following appendices is to illustrate some examples of the management approach and management tools used in practice.

These examples cover the following:

1. Organisation structure
2. Management approach
3. Control system
4. Control tools
5. Training

APPENDIX A

Organisation

Three organisational structure charts have been selected to illustrate different types of organisation. The type of organisation changes depend on the functions to be served.

Fig. A–1. The Capital Development Board Organisation (Illinois, USA) Chart, shows staff divisions and sections.

The C.D.B. provides technical or management services in three main areas:

1. Development of capital improvement programmes.
2. Capital budget enactment.
3. Project design and construction.

In programme development the agency works as a technical consultant to other state agencies or local governments in their building programmes. It assesses client agencies' building needs and helps them establish space requirements and dollar estimates for proposed projects.

C.D.B. also assists the Governor and Legislature in preparation and enactment of the annual capital budget.[1]

Fig. A–2. A typical functional organisation chart illustrates a management group which could be established by Holmes & Narver Inc. (Engineers – Constructors), USA, for a multi-site construction programme. The chart includes the organisation and the major tasks to be performed by each manager.[2]

Fig. A–3. Saudi Consulting House Organisation Structure. The objective of this agency is to provide different consulting services to Government agencies, private establishments and individuals including preparation of different research studies In addition, S.C.H. provides technical advice to existing industries and determines the economic feasibility for all types of projects as well as their priorities and methods of financing.[3]

1 'Capital Development Board: Philosophy Mission and Role', The Capital Development Board, Illinois, USA, 1977.
2 'Construction Management Procurement', Holmes & Narver Inc., Orange, California, USA, 1981.
3 'Saudi Consultant House Objectives', Saudi Consultant House, Riyadh, Saudi Arabia, 1982.

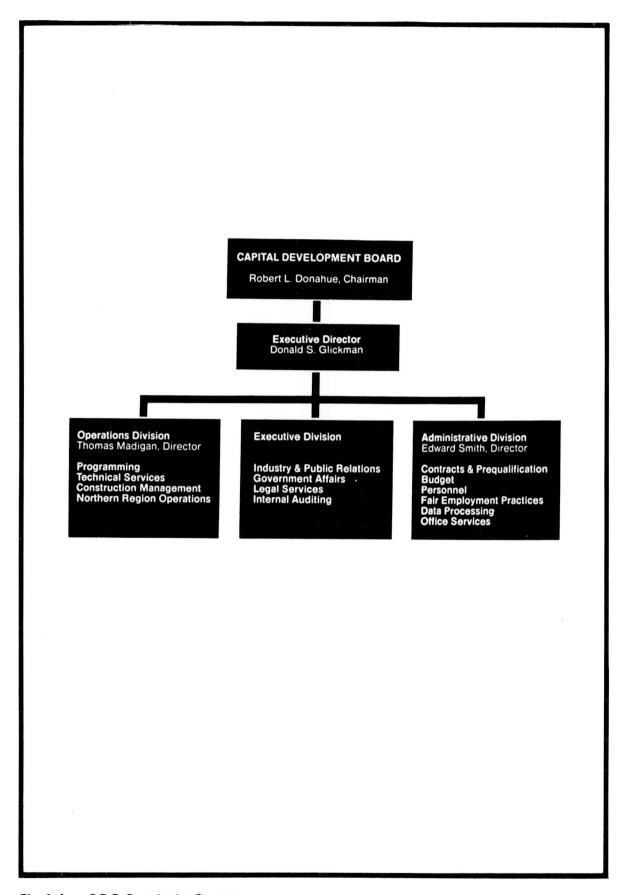

CAPITAL DEVELOPMENT BOARD

Robert L. Donahue, Chairman

Executive Director
Donald S. Glickman

Operations Division
Thomas Madigan, Director

Programming
Technical Services
Construction Management
Northern Region Operations

Executive Division

Industry & Public Relations
Government Affairs
Legal Services
Internal Auditing

Administrative Division
Edward Smith, Director

Contracts & Prequalification
Budget
Personnel
Fair Employment Practices
Data Processing
Office Services

Fig. A–1   C.D.B. Organisation Structure.
Source   The Capital Development Board, Illinois, USA.

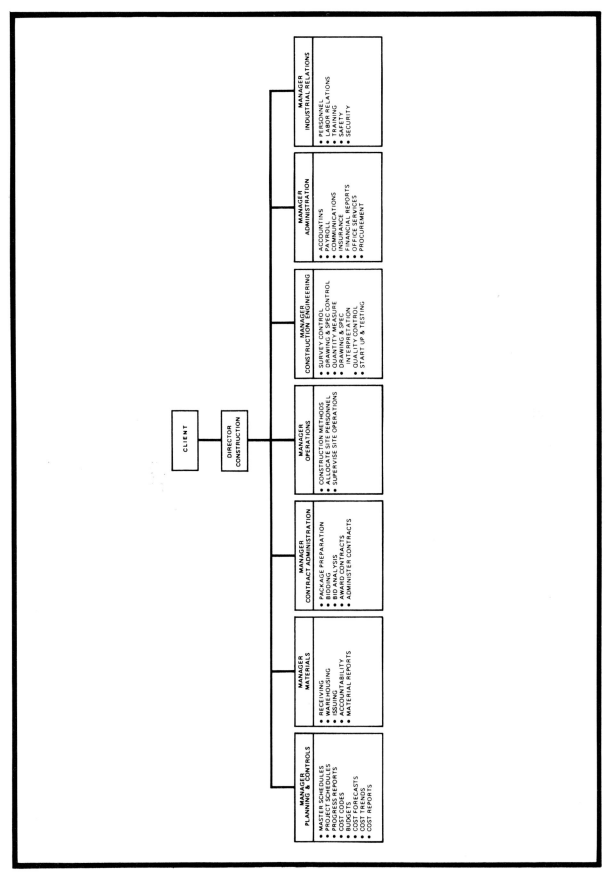

Fig. A–2   Functional Organisation Chart.

Source   Holmes & Narver Inc., Orange, California, USA.

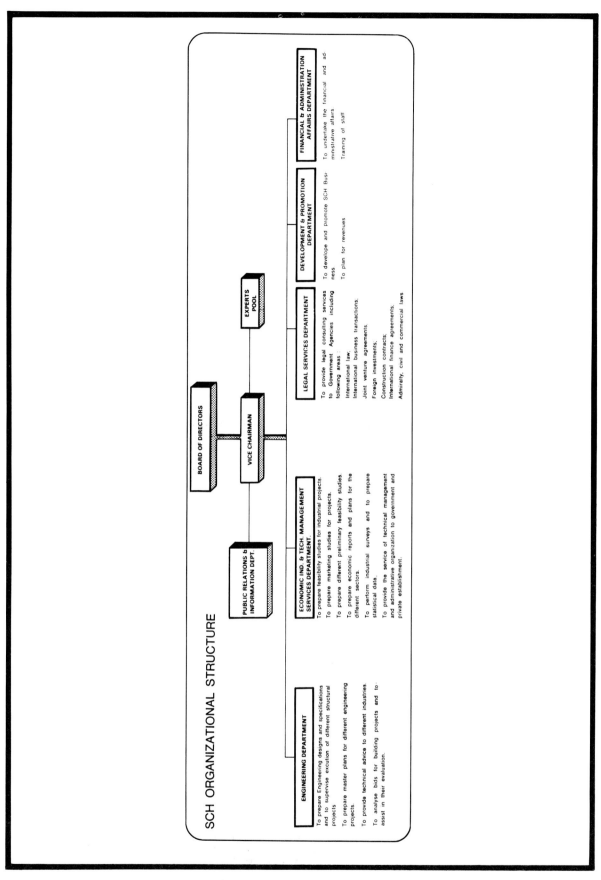

Fig. A–3 Saudi Consulting House Organisation Structure.
Source Saudi Consulting House, Riyadh, Saudi Arabia.

67

APPENDIX B

Approach and Methodology

Fig. B–1. The matrix contains a brief description of how the Arthur D. Little Programme Systems Management Company (USA) addresses the principal project planning elements.[1]

There are five key elements to managing large and complex projects effectively. They are: planning, organisation, control, training and implementation.

Fig. B–2. The chart illustrates the Development Process, as adopted by C.D.B., Illinois, USA.

Several steps have been included in the development of a new or improved public facility. This chart traces the procession outline starting with the user's agency needing the facility; it indicates the role and interaction of C.D.B., the Government's office, the Legislature and the design and construction firms engaged by C.D.B.[2]

Fig. B–3. Management Decision Network for a Large Scale Project adopted by Gibbs & Hill Inc. (A/E) from USA is illustrated.

Gibbs & Hill offers Management Services to cover every area of activity from project inception to full-scale operations including:

– Development of client corporate goals and objectives, study designs and management decision networks.
– Programming and scheduling project related activities and decisions of client's corporate management.
– Economic planning, including development schedules for capital funds, revenue and expense projections and project feasibility analysis.
– Development and evaluation of alternative project objectives, methods, processes and schedules.
– Selection and supervision of spatial consultants, contractors, etc.
– Site search and selection.
– Project design.
– Construction.
– Project staffing and manpower training.
– Start-up operation.[3]

Fig. B–4. The chart illustrates the engineering approach adopted by Chas. T. Main Inc., USA (A/E).

Chas. T. Main is active in capital programmes offering each of these areas comprehensive professional engineering services involving modernisation, addition and new facilities. Assignments in these fields range from feasibility studies through project planning, final design and construction management.[4]

1 'Elements of Project Management', Arthur D. Little Program Systems Management Company, Cambridge, USA, 1982.
2 'Capital Development Board: Philosophy, Mission and Role'. The Capital Development Board, Illinois, USA, 1977.
3 'Management', Gibbs & Hill Inc., New York, USA, 1980.
4 'Planning and Controls', Holmes & Narver Inc., Orange, California, USA, 1981.

Principal Project Planning Elements

Planning System \ Planning Stage	Feasibility	Project Strategy	Baseline Plan	Detailed Planning
Economic Evaluation	• Benefits • Risk	• Continue Appraisal with View to Changing Project Specifications if Necessary	• Impact on other business functions assessed • Adjustments made as necessary	
Project Definition	• System Specs • Base Technology • $ Estimate • Project Schedule	• Outline Design • Configuration Definition • Budget by Major Areas • Milestone Schedule • Detailed "Planning" Schedule	• Further Development of Outline Design, Schedule and Budget	• Detailed Contract Specs and Drawings • Overall Schedule Requirements • Detailed Budget/Contract Bids
Finance	• Potential Sources	• Principal Sources • Major Payments	• Detailed Sources • More Detailed Cash Requirements	• Detailed Payments Schedule by Creditor & Currency
Environment	• Initial Impact Assessment	• Definition of Environmental Impact Statement • National & Local Government Support Assessed • Local Population Attitude Assessed • Supplier Situation Assessed	• Schedule of Approvals Required • Government or Community Support Groups Identified	• Permit Expediting System • Expediting Schedules • Public Relations
Organization & Systems	• Initial Project Outline	• Overall Concepts for: —Contractor Strategies —Design, Fabrication, Construction —Labor & Materials Sources • Principal Responsibilities Determined • Major Information Systems Identified • Key Personnel Identified	• Some Major Contracts Signed • Union Discussions • Possibly Some Long Lead Materials Ordered • Responsibilities Matrix • Manpower Plan • Systems Design Schedule	• Contract Terms and Conditions • Owner Organization Detailed • Detailed Staffing Plans
Infrastructure & Support	• Assessment Extent of Support Required	• Preliminary Plans for: —Labor Relations —Camps —Logistics	• Further Definition of: —Labor Relations —Camps —Administration —Transport, Logistics & Warehousing • Support Organization Outlined • Permits Requested Outlined	• Detailed Definition of: —Labor Relations —Camps —Transport, Logistics, etc. • Construction Schedules/Contracts for Camps, Power, Transport, etc. • Service Contracts Identified • Support Organization Defined

Fig. B–1   Principal Project Planning Elements.
Source   Program Systems Management Company, Cambridge, Massachusetts, USA.

Program Development

1 User agency identifies a need for new or improved facility	2 Evaluation and cost estimate by CDB	3 Project scopes prepared by User and CDB
4 Capital Budget Request prepared by User	5 Governor's Capital Program prepared	6 Appropriation Legislation drafted by CDB and Bureau of the Budget

Capital Budget Enactment

7 Appropriation Bills introduced	8 Information to legislative staff by CDB and User	9 Passage and signature by Governor

Project Implementation

10 Requests for release of funds by User and CDB	11 Governor's release of funds	12 Architect/Engineer selected by User and CDB
13 Design phase CDB and User agency approve A/E design for construction	14 Advertisement, bidding and award of construction contracts by CDB	15 Construction phase CDB monitors work of A/E and contractor through construction
16 Transfer of completed project from CDB to User		

Fig. B–2   Development Process.
Source   The Capital Development Board, Illinois, USA.

70

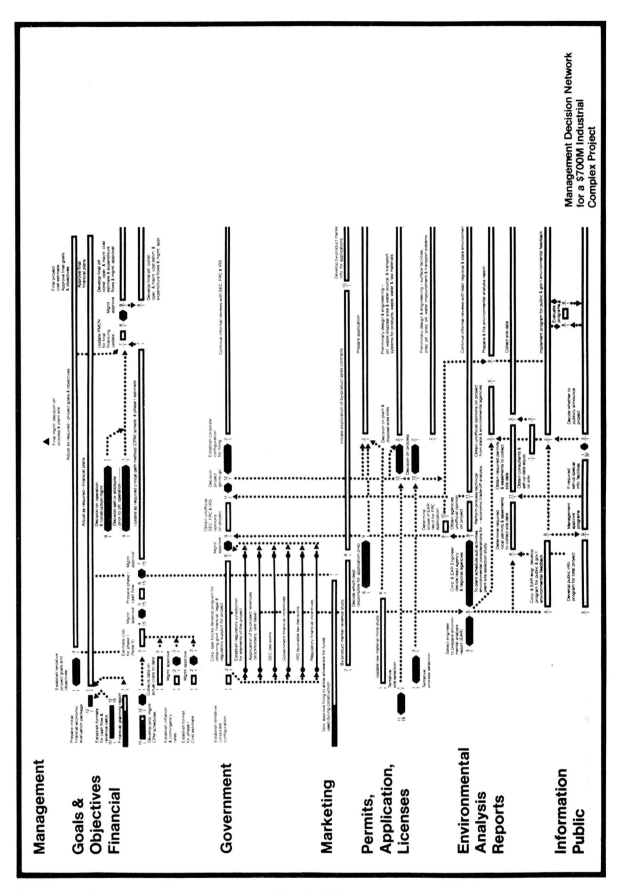

Fig. B–3   Management Decision Network for a Complex Project.
Source   Gibbs & Hill Inc., New York, USA.

71

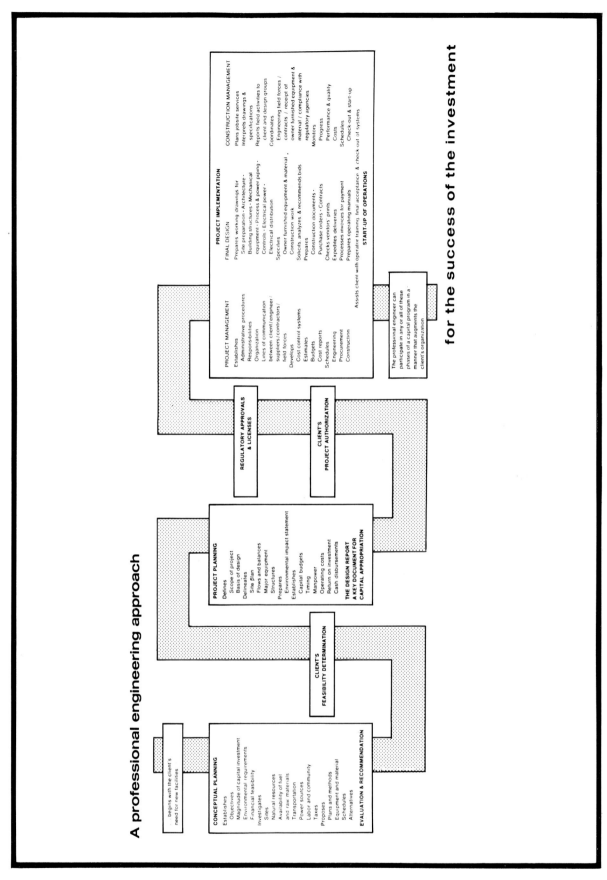

Fig. B–4   Engineering Approach.
Source   Chas. T. Main Inc., Boston, Massachusetts, USA.

72

APPENDIX C

Control system

Fig. C–1. Holmes & Narver's project planning system encompasses:

1. A planned approach.
2. A cost control base.
3. Accurate and timely reporting of expenditures and work progress.
4. An alert system for cost and schedule impact resulting from deviations.
5. Positive action to correct deviations.
6. Active participation by all project members.

The planned approach relates the scope of work to the schedule and cost estimate: it is one of the first steps in project management and begins with the initiation of the work. An early schedule and an early budget become the control base from which Holmes & Narver works. This schedule and budget for the project are refined as the work progresses.

The Project Manager is directly responsible for controlling the cost and the schedule of the project and furnishes the positive action to correct deviations from the original plan that are brought to his attention by the control tools furnished within the system.[1]

The diagram of the Holmes & Narver Planning and Control System illustrates the inter-relationships of the various activities, controls and reports. The planning and scheduling proceeds concurrently with the cost estimating. The procedure for planning is shown on the upper portion of the diagram, while the cost estimate developments are outlined on the lower portion. Throughout the project the importance of this relationship is recognised.

Fig. C–2. The Parsons Project Control System consists of three separately identifiable but related functions.

1. Development of a project estimate and budget.
2. Establishment and maintenance of a cost control system.
3. Development of a work plan and schedule.

Project Control, as practised by the Ralph M. Parsons Company, is a systematic, continuing activity of recording and forecasting incremental and overall project costs and progress, and comparing such costs and progress against the project budget and schedule during job accomplishment. This system has a two-fold objective.

Firstly, the system controls cost and progress against a predetermined budget and schedule. Secondly, it provides the Parsons Management with timely and accurate cost and progress status information on both a current and a forecast-at-completion basis.

The Parsons Project Control System is built upon the following key principles:

1. An estimate (budget) and project plan (schedule) must be prepared early in a project cycle.
2. Costs and progress must be continuously compared with the budget and the schedule. This comparison must be applied during all project phases; i.e., design, procurement and construction.
3. Project costs and schedule performance must be reported, forecasted and controlled.
4. Established Parsons Project Control procedures will be rigidly enforced.
5. Cost and schedule control are continuing responsibilities of management and the entire project team.

The organisational relationship of the Project Manager to the Project Control Manager is depicted in Fig. C–2.[2]

Fig. C–3. This system is designed to be used by a British contractor for a large scale project. The areas of site activity covered by the system are:

1. Network planning – project controls, using time and resource analysis linked to material and cost files in the data base.
3. Material control – through linking drawing registers with material take-off and stock control to perform material allocation.
3. Cost control – time sheet, productivity and original bid analysis procedures.
4. Cost forecasting – monitoring original bid estimates with actual and forecast costs.

It is an inter-active prototype Management Control System based on a model project. This system forms the basis for establishing computer procedures on a live site by making a copy of the prototype system and tailoring it by means identified, modifications and enhancements.[3]

---

1. 'Professional Engineering for Capital Programs', Chas. T. Main Inc., Boston, USA, 1981.
2 'Project Control System', The Ralph M. Parsons Company, USA, 1980.
3 'Outline Design of an Interactive Management Control System', Computing Services, Borehamwood, UK, 1979.

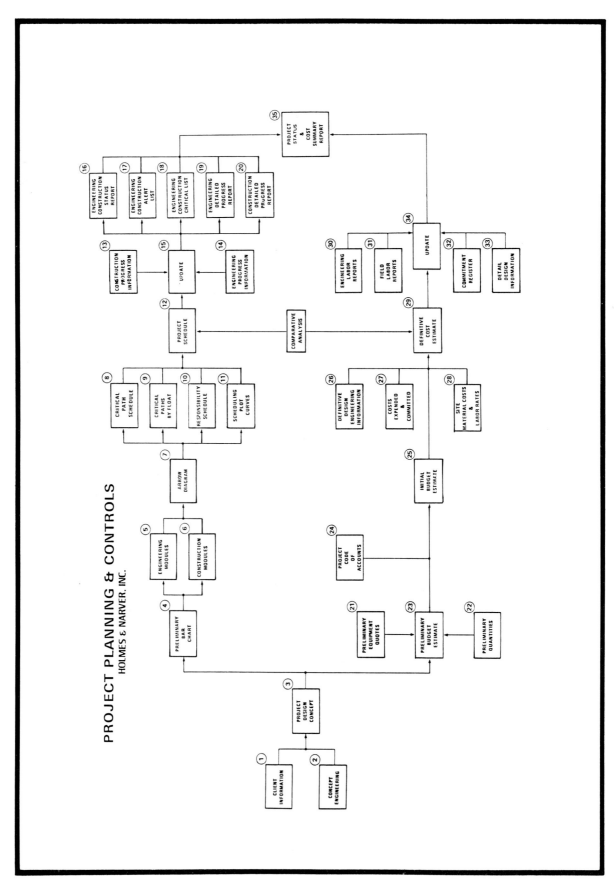

Fig. C–1   Project Planning & Controls.

Source   Holmes & Narver Inc., Orange, California, USA.

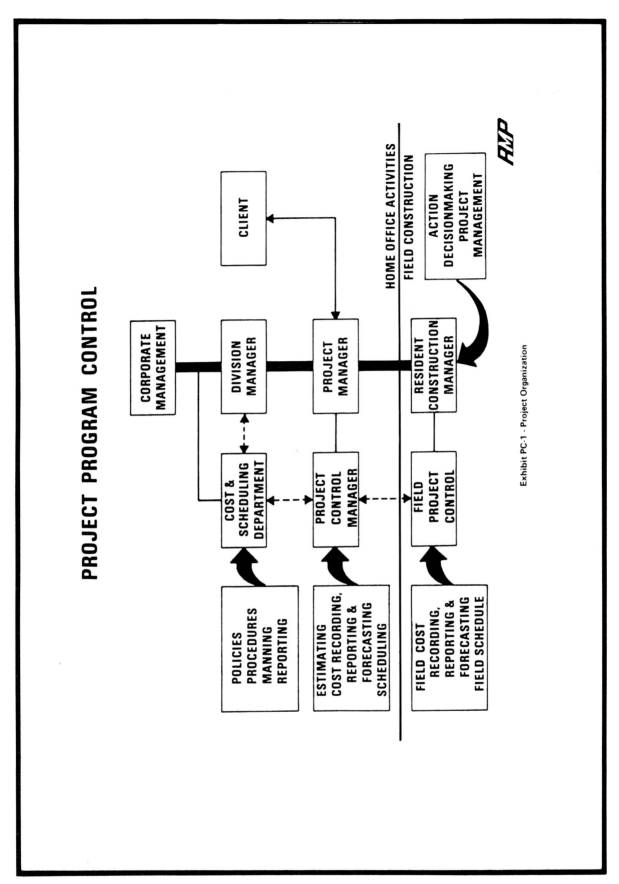

PROJECT PROGRAM CONTROL

Exhibit PC-1 - Project Organization

Fig. C–2 Project Program Control.
Source The Ralph M. Parsons Company, USA.

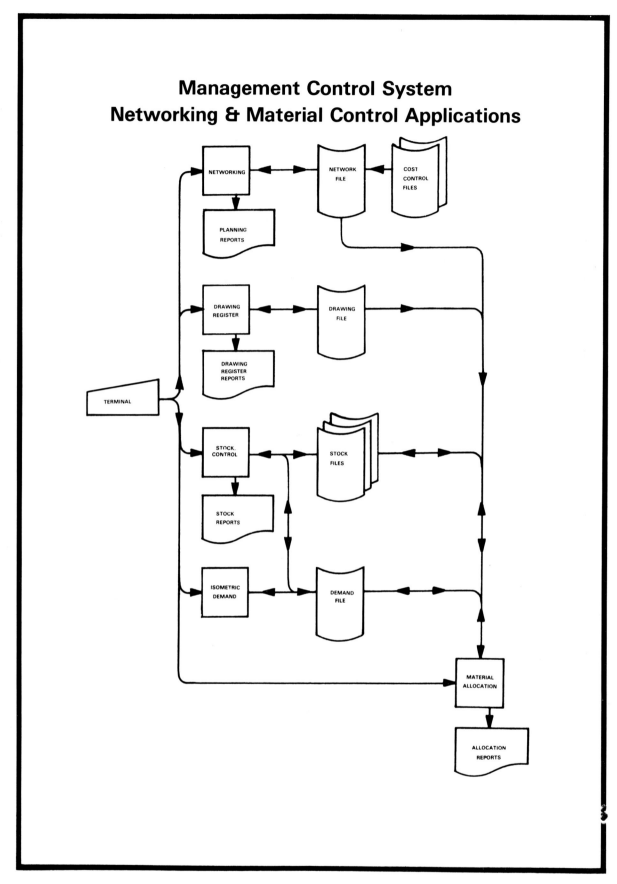

Fig. C–3/1   Management Control System Networking & Material Control Applications.
Source   Computing Services, Borehamwood, UK.

Management Control System
Cost Control & Cost Forecasting Applications

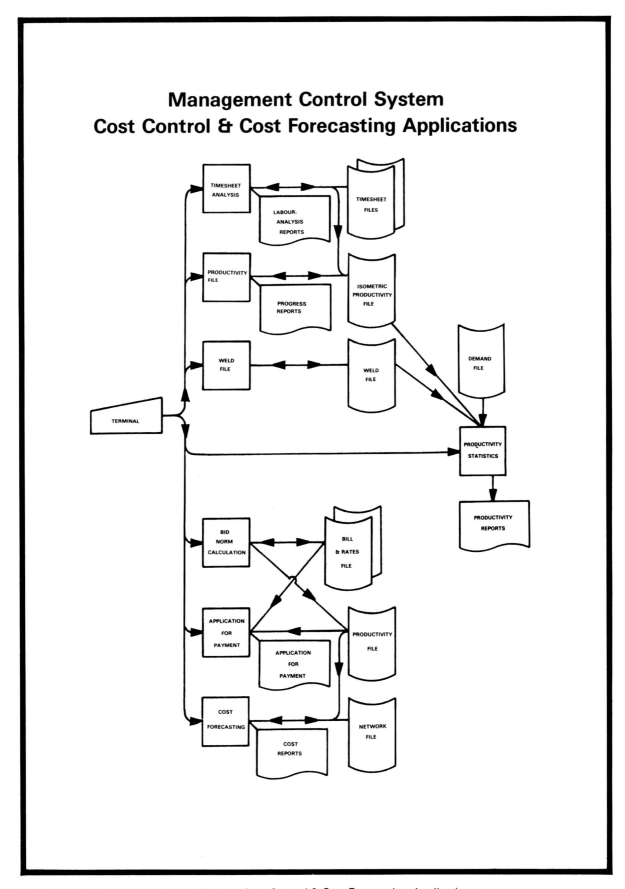

Fig. C–3/2 Management Control System Cost Control & Cost Forecasting Applications.
Source Computing Services, Borehamwood, UK.

77

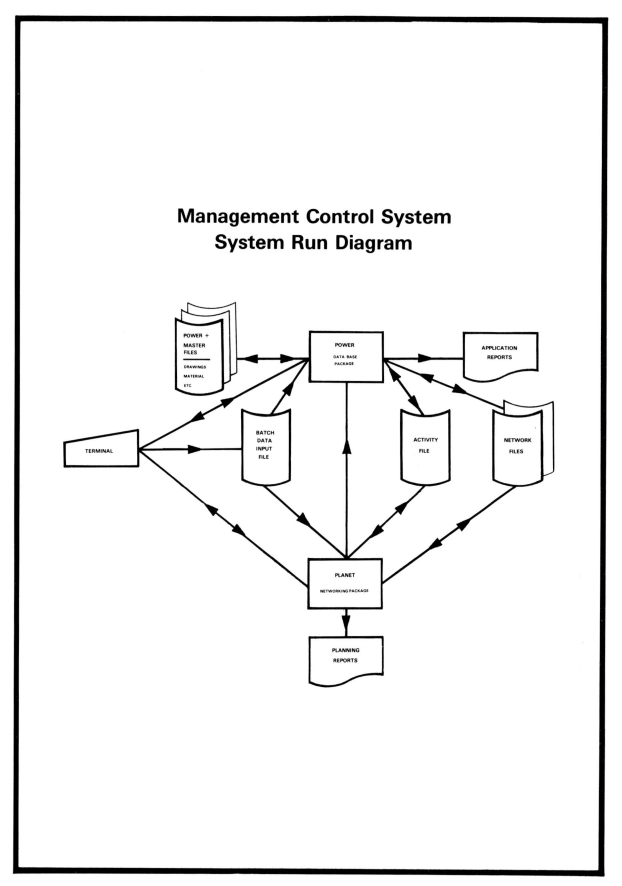

Fig. C–3/3 Management Control System System Run Diagram.
Source Computing Services, Borehamwood, UK.

APPENDIX D

Planning and Scheduling Control tools

The tools illustrated and the definitions given below are reproduced from the Holmes & Narver Inc. Control System.

Fig. D–1. Arrow Diagram. Once the logic of the engineering and the construction modules have been selected the activities are arranged in sequence and they are combined on a project arrow diagram. A portion of such a diagram (much reduced) is shown in Fig. D–1.

Fig. D–2. Critical Path Schedule. Upon the completion of the arrow diagram and the assignment of duration to all activities, the schedule is computed using a computer program. The first output of this program is the Critical Path Schedule which lists every activity in the Project Performance Plan. Other outputs of the program form the complete initial schedule, and illustrations of these outputs are shown on the next figures.

Fig. D–3. Critical Paths by Float. This output lists activities, in ascending float order, which affect the project completion. (Float is the time an activity can be delayed without affecting the completion of the project.)

Fig. D–4. Responsibility Schedule. The responsibility schedule is a computer-drawn bar-chart showing activities in early start order by each supervisor's responsibility.

Fig. D–5. Scheduling Plot Curves. These curves are produced by the computer and they show the schedule percentages of completion at various times on two bases: the first basis shown by a line on the left shows that all activities are performed at the earliest starting date; whereas the second curve indicates that all activities are performed on the latest starting date.

Fig. D–6. Project Schedule. The prededing reports are analysed and compared with the cost estimates being prepared simultaneously with the scheduling and with the various manpower requirements. They are revised as necessary, and, finally, the 'best plan' is selected which becomes the program performance plan. At this stage in the system an arrow diagram of the 'best plan' is produced by the computer.

Fig. D–7. Engineering – Construction Status Report. This shows all progress on started activities and indicates activities that did not start on the scheduled date.

Fig. D–8. Engineering – Construction Alert List. Details items on status reports that are within 20 days of becoming critical to project completion. The list is directed to the specific attention of the Project Manager for review and necessary action.

Fig. D–9. Engineering – Construction Critical Activities List. Shows all activities within 10 days of becoming critical to project completion. The list is directed to the specific attention of the Project Manager for review and action.

Fig. D–10. Engineering Detailed Progress Report. The progress of the engineering activities, including status of drawings and the estimated hours to complete each category, is summarised and presented as shown in the Engineering Progress Report.

Fig. D–11. Construction Detailed Progress Report. The various construction progress information reports are allocated to the various construction functions and are summarised as shown in the Construction Progress Report.

Fig. D–12. Project Status Report Schedule. The reports mentioned above are reviewed by the Project Manager and become the Project Status Report. They are summarised graphically as shown in Fig. D–12.

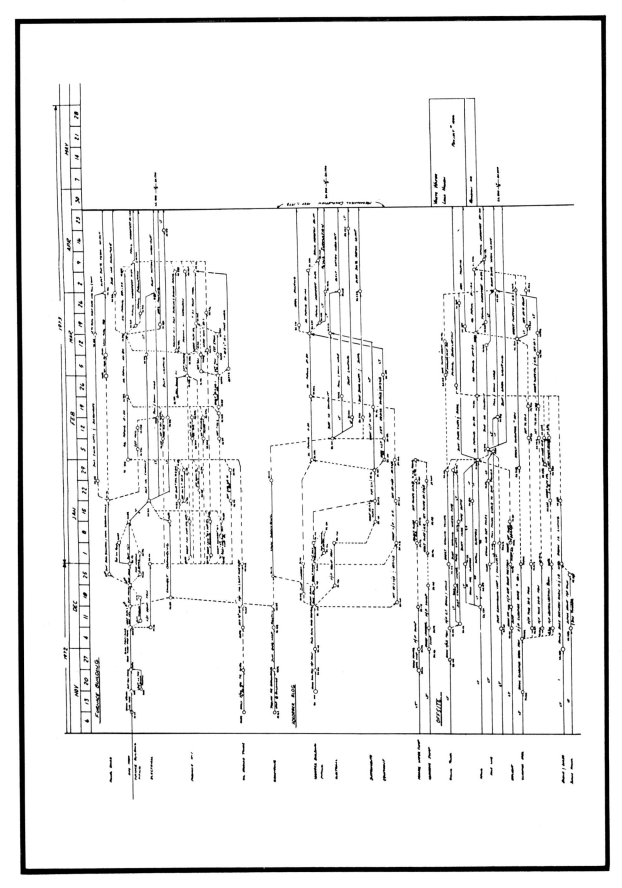

Fig. D–1 Arrow Diagram.
Source Holmes & Narver Inc., Orange, California, USA.

CONTRACT - 1200
AREA - 100
CUSTOMER - WASTE WATER

LOCK HAVEN

CRITICAL PATH DETAIL LISTING

B050010

ORIG. START DATE 13NOV72
AS OF DATE 13NOV72
PAGE 1

C P NODE	I NODE	J NODE	DURA-TION	CAL NO	ACCT NUMBER	R S P	PCT COMPT	ACTIVITY DESCRIPTION	EARLY START	LATE START	EARLY FINISH	LATE FINISH	TOTAL FLT	FREE FLT
AA100	AA102		5	01	000	2	0	FB FUR BLDG EXCAV FDN	N13NOV72	13NOV72	7NOV72	21NOV72	2	0
AA102	AA104		0	01	000	0	0	FB DUMMY	20NOV72	22NOV72	17NOV72	21NOV72	2	0
AA102	AA106		5	01	100	2	0	FB FUR BLDG INST FDN FORMS	20NOV72	22NOV72	24NOV72	28NOV72	2	0
AA104	AA106		5	01	500	2	0	FB FUR BLDG INST UG PLUMBING&PIPE	20NOV72	22NOV72	24NOV72	28NOV72	2	0
AA106	AA108		10	01	100	2	0	FB FUR BLDG POUR FDN-SLAB-STRIP FORM	27NOV72	29NOV72	8DEC72	12DEC72	2	0
AA108	AA110		2	01	000	0	0	FR LEADTIME	11DEC72	14DEC72	12DEC72	15DEC72	3	0
AA108	AA114		0	01	000	0	0	FB DUMMY	11DEC72	20DEC72	8DEC72	19DEC72	7	0
AA108	AA118		5	01	300	2	0	FB FUR BLDG ERECT BLDG&ROOF(STEEL:	11DEC72	13DEC72	15DEC72	19DEC72	2	0
AA110	AA112		7	01	300	2	0	FH FUR BLDG MASONRY WORK	13DEC72	13DEC72	21DEC72	26DEC72	3	0
AA112	AA120		0	01	000	0	0	FB DUMMY	22DEC72	27DEC72	21DEC72	26DEC72	3	1
AA114	AA116		15	01	100	2	0	FB EQUIP FDN F&F	11DEC72	20DEC72	29DEC72	9JAN73	7	0
AA116	AA170		0	01	000	0	0	FB DUMMY	1JAN73	10JAN73	29DEC72	9JAN73	7	0
AA116	AA178		0	01	000	0	0	FH DUMMY	1JAN73	10JAN73	29DEC72	9JAN73	7	0
AA116	AA178		0	01	000	0	0	FH DUMMY	1JAN73	10JAN73	29DEC72	9JAN73	7	0
AA116	AA182		0	01	000	0	0	FB DUMMY	1JAN73	24JAN73	29DEC72	23JAN73	17	13
AA116	AA186		0	01	000	0	0	FB DUMMY	1JAN73	29JAN73	29DEC72	26JAN73	20	13
AA116	AA190		0	01	000	0	0	FB DUMMY	1JAN73	31JAN73	29DEC72	30JAN73	22	15
AA116	AA194		0	01	000	0	0	FB DUMMY	1JAN73	31JAN73	29DEC72	30JAN73	22	15
AA116	AA200		0	01	000	0	0	FB DUMMY	1JAN73	21FEB73	29DEC72	20FEB73	37	30
AA116	AA204		0	01	000	0	0	FB DUMMY	1JAN73	21FEB73	29DEC72	20FEB73	37	30
AA116	AA230		0	01	000	0	0	FB DUMMY	1JAN73	21FEB73	29DEC72	20FEB73	37	30
AA118	AA120		5	01	300	2	0	FB FUR BLDG INST SIDING	18DEC72	20DEC72	22DEC72	26DEC72	2	0

Fig. D–2 Critical Path Schedule.

Source Holmes & Narver Inc., Orange, California, USA.

CONTRACT - 1200
AREA 100 MASTE MTR
MASTE MATER

LOCK HAVEN

CRITICAL PATH DETAIL LISTING IN AREA/FLOAT/I/J SEQUENCE

C P NODE	I NODE	J NODE	DURA-TION	CAL NO NUMBER	ACCT NUMBER	R S P	PCT COMPT	ACTIVITY DESCRIPTION	EARLY START	LATE START	EARLY FINISH	LATE FINISH	TOTAL FLT	FREE FLT
*	AA250	AA252	5		400	2	0	FB DELIVER FURNACE 11-1	N26FEB73	26FEB73	2MAR73	2MAR73	0	0
*	AA252	AA254	5		400	2	0	FB FURNACE 11-1 SET	5MAR73	5MAR73	9MAR73	9MAR73	0	0
*	AA254	AA262	0		000	0	0	FB DUMMY	12MAR73	12MAR73	9MAR73	9MAR73	0	0
*	AA262	AA264	5		400	2	0	FB FURNACE 11-1 ERECT PLATF STEEL	12MAR73	12MAR73	16MAR73	16MAR73	0	0
*	AA264	AA266	15		400	2	0	FB FURNACE 11-1 DUCT WORK	19MAR73	19MAR73	6APR73	6APR73	0	0
*	AA266	AA366	2		000	0	0	FB LEADTIME	9APR73	9APR73	10APR73	10APR73	0	0
*	AA366	AA500	15		500	2	0	FB PIPING HYDROTEST 50-100	11APR73	11APR73	1MAY73	1MAY73	0	0
*	AA500	DA500	0		000	0	0	FB LEADTIME	2MAY73	2MAY73	1MAY73	1MAY73	0	0
*	CA208	CA260	30		400	2	0	OFF CLARIFIER ERECT	N 5FEB73	5FEB73	16MAR73	16MAR73	0	0
*	CA260	CA262	15		200	2	0	UFF CLARIFIER ERECT TOP PLATF&HR	19MAR73	19MAR73	6APR73	6APR73	0	0
*	CA262	CA326	2		000	0	0	OFF DUMMY	9APR73	9APR73	10APR73	10APR73	0	0
*	CA326	CA346	0		000	0	0	OFF DUMMY	11APR73	11APR73	10APR73	10APR73	0	0
*	CA346	CA500	15		500	2	0	OFF PIPING HYDROTEST 50-100	11APR73	11APR73	1MAY73	1MAY73	0	0
*	CA500	DA500	0		000	0	0	OFF LEADTIME	2MAY73	2MAY73	1MAY73	1MAY73	0	0
	CA133	CA132	5		000	0	0	OFF LEADTIME	N19DEC72	20DEC72	25DEC72	26DEC72	1	0
	CA130	CA136	25		000	2	0	UFF SLEEPERS EXCAV	N19DEC72	20DEC72	22JAN73	23JAN73	1	0
	CA132	CA300	30		100	2	0	OFF SLEEPERS INSTALL	26DEC72	27DEC72	5FEB73	6FEB73	1	0
	CA136	CA300	10		000	0	0	OFF LEADTIME	23JAN73	24JAN73	5FEB73	6FEB73	1	0
	CA300	CA316	15		500	2	0	OFF AG PIPING 0-40	6FEB73	7FEB73	26FEB73	27FEB73	1	0
	CA300	CA360	40		700	2	0	OFF INSTR SUPTS&INSTR INSTALL	6FEB73	7FEB73	2APR73	3APR73	1	0
	CA310	CA500	20		600	2	0	OFF ELECT&INSTR SYSTEM CK-OUT	3APR73	4APR73	30APR73	1MAY73	1	1
	CA316	CA320	15		500	2	0	OFF AG PIPING 40-80	27FEB73	28FEB73	19MAR73	20MAR73	1	0
	CA316	CA356	0		000	0	0	OFF DUMMY	27FEB73	28FEB73	26FEB73	27FEB73	1	0

Fig. D–3 Critical Paths by Float.
Source Holmes & Narver Inc., Orange, California, USA.

82

SCHEDULE BAR CHART

CONTRACT - 1206
AREA - 100
STATUS DATE - 13NOV72

SORT OPTION - 2 CONTRACT,ACCOUNT,EARLY START,I-NODE,J-NODE

B050120
PAGE - 0001
PART - 01

ALL DATES USED ARE CPM CALCULATED

I NODE	J NODE	CL NO	ACCT NUMBER R	ACTIVITY DESCRIPTION		PCT COMP	REM TOTL PLN
AA100	AA102	01	000 2	FB FUR BLDG EXCAV FDN	A	0	2
AA154	AA156	01	000 2	FB OIL STG 1434-5 EXCAV	B	0	37
AA164	AA166	01	000 2	FB TRENCH GROUND LOOP AT BUILDINGS	C	0	7
BA100	BA102	01	000 2	AB ADS BLDG EXCAV FDN	D	0	2
CA200	CA202	01	000 2	OFFI CLARIFIER AREA EXCAV FDN	E	0	20
BA170	BA172	01	000 2	AB PROCESS MTR SUMP EXCAV	F	0	22
BA190	BA192	01	000 2	AB ADSORBER SUMP EXCAV	G	0	14
CA100	CA102	01	000 2	OFFI COOLING TOWER EXCAV FDN	H	0	2
CA280	CA286	01	000 2	OFFI BUILD DIKE BETWEEN BASIN 1A&1R	I	0	27
AA168	AA169	01	000 2	FB GROUNDING LOOP TRENCH BACKFILL	J	0	7
CA170	CA172	01	000 2	OFFI EFFLUENT SUMP EXCAV	K	0	17
CA102	CA104	01	000 2	OFFI COOLING TOWER F&P BASINWALLS	L	0	2
CA130	CA136	01	000 2	OFFI SLEEPERS EXCAV	M	0	1
AA102	AA106	01	100 2	FB FUR BLDG 1VST FDN FORMS	N	0	2
AA106	AA108	01	100 2	FB FUR BLDG POUR FDN-SLAB-STRIP FORM	O	0	2
BA102	BA104	01	100 2	AB ADS BLDG F&P FDN	P	0	2
BA104	BA106	01	100 2	AB ADS BLDG F&P SLAB	Q	0	2
BA172	BA174	01	100 2	AB PROCESS MTR SUMP F&P SUMP	R	0	22
AA114	AA116	01	100 2	FB EQUIP FDN F&P	S	0	7
BA192	BA194	01	100 2	AB ADSORBER SUMP F&P SUMP	T	0	14
CA202	CA208	01	100 2	OFFI CLARIFIER BASE F&P	U	0	20
CA210	CA212	01	100 2	OFFI TANK 34-4 F&P FDN	V	0	37
CA220	CA222	01	100 2	OFFI TANK 34-10 F&P FDN	W	0	38
CA280	CA282	01	100 2	OFFI NEUTRALIZATION BASIN F&P	X	0	27
CA282	CA284	01	100 2	OFFI INTAKE STR F&P BASIN&INST TIMBER	Y	0	30

Fig. D–4 Responsibility Schedule.

Source Holmes & Narver Inc., Orange, California, USA.

83

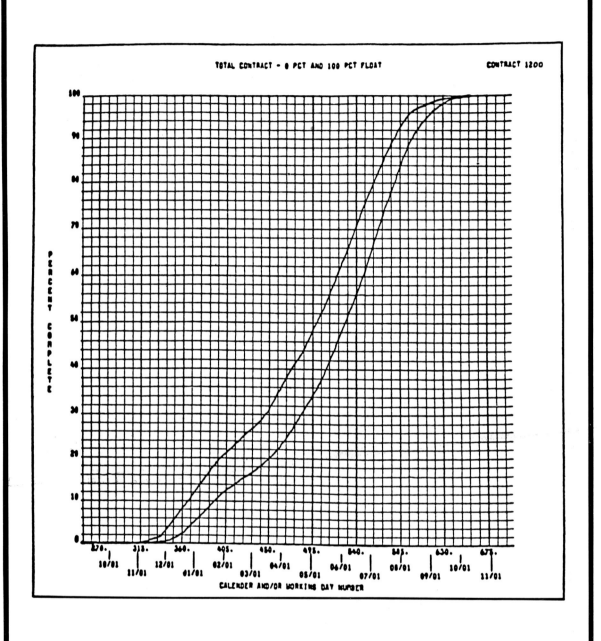

Fig. D–5 Scheduling Plot Curves.
Source Holmes & Narver Inc., Orange, California, USA.

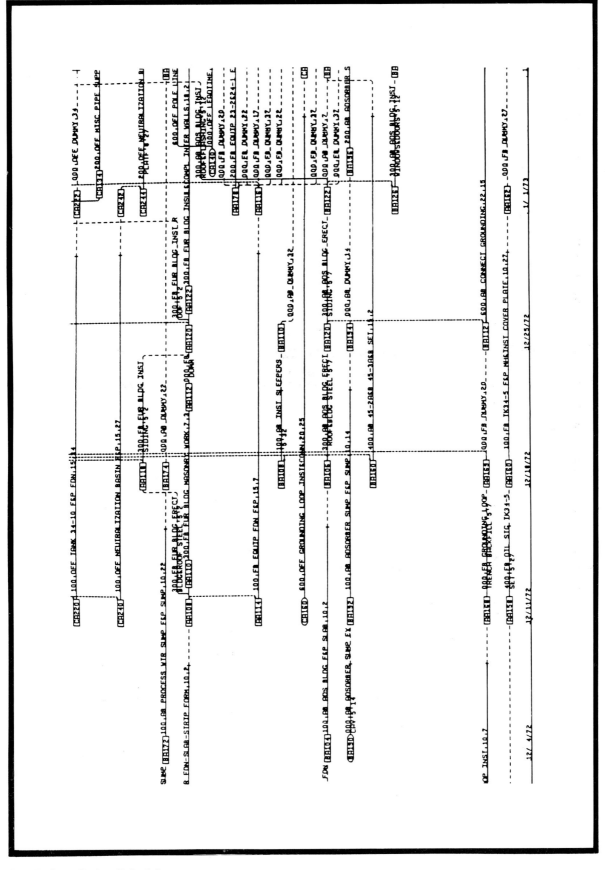

Fig. D–6 Project Schedule.

Source Holmes & Narver Inc., Orange, California, USA.

CONSTRUCTION STATUS REPORT

REPORT STATUS AS OF 13 NOV 72
8050050

WASTE WATER
LOCK HAVEN

RECORDS ARE INCLUDED IN REPORT IF REM FLOAT IS LESS OR EQUAL
9999 DAYS AND IF SCHED START DATE IS WITHIN 99 DAYS OF REPORT DATE
SORT #3 CONTRACT, ACCOUNT, SCHEDULED START

PAGE 1
DATE USER SUPPLIED

AREA	ACCT	RES	I	J	DUR	PCT COMP	ACTIVITY DESCRIPTION	SCHED START D M Y	EST START D M Y	ACT START D M Y	EST COMPL D M Y	FLOAT TOT REM	CRD CDE	ALERT	RPT MGR
100	000	2	AA100	AA102	0005	000	FB FUR BLDG EXCAV FDN	131172			171172	2 2	C53	X	->
100	000	2	AA154	AA156	0010	000	FB OIL STG TK34-5 EXCAV	131172			241172	37 37	C53		
100	000	2	AA164	AA166	0010	000	FB TRENCH GROUND LOOP AT BUILDINGS	131172			281172	7 7	C53	X	->
100	100	2	AA102	AA106	0005	000	FB FUR BLDG INST FDN FORMS	201172				2 2	C83	X	->
100	500	2	AA104	AA106	0005	000	FB FUR BLDG INST UG PLUMBING&PIPE	201172				2 2	C53	X	->
100	000	2	BA100	BA102	0005	000	AB ADS BLDG EXCAV FDN	201172				2 2	C53	X	->
100	000	2	CA200	CA202	0015	000	OFF CLARIFIER AREA EXCAV FDN	201172				20 20	C53	X	
100	100	2	AA106	AA108	0010	000	FB FUR BLDG POUR FDN-SLAB-STRIP, FORM	271172				2 2	C53	X	->
100	630	2	AA166	AA168	0010	000	FB GROUNDING LOOP INST	271172				7 7	C53	X	->
100	100	2	BA102	BA134	0005	000	AB ADS BLDG F&P FDN	271172				2 2	C53	X	->
100	000	2	BA170	BA172	0005	000	AB PROCESS MTR SUMP EXCAV	271172				22 22	C53	X	
100	100	2	BA104	BA106	0010	000	AB ADS BLDG F&P SLAB	041272				2 2	C53	X	->
100	100	2	BA172	BA174	0010	000	AB PROCESS RTR SUMP F&P SUMP	041272				22 22	C53	X	
100	000	2	BA190	BA192	0005	000	AB ADSORBER SUMP EXCAV	041272				18 14	C53	X	
100	000	2	CA100	CA102	0008	000	OFF COOLING TOWER EXCAV FDN	041272				2 2	C53	X	->
100	000	2	CA280	CA286	0020	000	OFF BUILD DIKE BETWEEN BASIN 1&1B	041272				27 27	C53	X	
100	300	2	AA108	AA118	0005	000	FB FUR BLDG ERECT BLDG&ROOF STEEL	041272				2 2	C53	X	->
100	100	2	AA114	AA116	0015	000	FB EQUIP FDN F&R	111272				7 7	C53	X	->
100	400	2	AA158	AA160	0005	000	FB OIL STG TK34-5 SET	111272				27 27	C63	X	->
100	000	2	AA166	AA169	0005	000	FB GROUNDING LOOP TRENCH BACKFILL	111272				7 7	C63	X	->

Fig. D–7   Engineering – Construction Status Report.
Source   Holmes & Narver Inc., Orange, California, USA.

CONTRACT 1200

CONSTRUCTION ALERT ACTIVITIES LIST

REPORT STATUS AS OF 16 JUL 71
B050050

RECORDS ARE INCLUDED IN REPORT IF REM FLOAT IS LESS OR EQUAL 20 DAYS.

SORT #1 CONTRACT, AREA, ACCOUNT, SCHEDULED START

DATE USER SUPPLIED
PAGE 1

AREA	ACCT	RFS	I	J	DUR	PCT COMP	ACTIVITY DESCRIPTION	SCHED START	EST START	ACT START	EST COMPL	FLOAT TOT REM	CRD CDE
030	000000	2	3504	3505	0005	095	EXC TUNNEL	170571		200571	160771	13 -26	C53
030	000000	2	8003	3004	0020	090	EXC PUMP HSE	170571		240571	190771	0 -25	C53
030	000000	2	3505	3575	0005	090	EXC TUNNEL	240571		270671	160771	13 -21	C53
030	000000	2	3575	3576	0005	040	EXC TUNNEL	010671		070671	160771	13 -15	C53
030	000000	2	3576	3577	0005	070	EXC TUNNEL	040671		140671	190771	10 -12	C53
030	000000	1	3577	3578	0005	050	EXC TUNNEL	150671		210671	190771	10 -1	C53
030	000000	2	3007	3009	0002	000	BAC FILL INSIDE 50	140771			190771	10 16	C53
030	000000	2	3548	3549	0005	000	BACKFILL	040871			150671	10 16	C53
030	000000	1	3010	3012	0002	000	BAC FILL INSIDE 50	110871				10 -1	C53
030	000000	1	3011	3013	0010	000	BAC FILL OUTSIDE	110871				0 -8	C53
030	100000	0	3560	3561	0005	000	BACKFILL	110871				13 13	C53
030	100000	0	3506	3507	0010	095	TUNNEL SLAB	240571		210571	160771	20 13	C53
030	100000	0	3533	3534	0010	095	TUNNEL SLAB	010671		080671	160771	23 0	C53
030	100000	0	3507	3508	0015	095	TUNNEL WALLS	080671		140671	160771	20 15	C53
030	100000	0	3545	3546	0010	098	TUNNEL SLAB	080671			160771	10 -18	C53
030	100000	0	3004	3005	0010	050	CONC FTG 50	150671		210671	220771	0 -18	C53
030	100000	0	3534	3535	0015	097	TUNNEL WALLS	150671		220671	160771	23 -15	C53
030	100000	0	3557	3558	0010	045	TUNNEL SLAB	150671		210671	190771	13 -1	C53
030	100000	0	3546	3547	0015	055	TUNNEL WALLS	220671		280671	260771	10 9	C53
030	100000	0	3578	3579	0010	040	TUNNEL SLAB	220671		280671	230771	10 5	C53
030	100000	0	3006	3007	0015	000	CONC FTG 50	290671			290771	5 -7	C53
030	100000	0	3558	3559	0015	000	CONC WALLS 50	290671			050871	10 -12	C53
030	100000	0	3579	3580	0015	008	TUNNEL WALLS	070771			050871	13 -1	C53
030	100000	0	3547	3548	0015	000	TUNNEL ROOF	140771			050871	10 14	C53
030	100000	0	3008	3010	0015	000	CONC WALLS 50	210771				0 0	C53
030	100000	0	3559	3560	0015	000	TUNNEL ROOF	210771				13 13	C53
030	100000	0	3580	3581	0015	000	TUNNEL ROOF	280771				10 10	C53
030	100000	0	3549	3550	0020	000	RING WALLS	110871				10 10	C53
030	100000	0	3012	3014	0008	000	CONC GRD SLAB	130871				0 0	C53
030	500000	7	3605	3606	0060	000	TAILINGS LINE	170571	010672		230872	269 6	C53

Fig. D–8 Engineering – Construction Alert List.

Source Holmes & Narver Inc., Orange, California, USA.

87

CONSTRUCTION CRITICAL ACTIVITIES LIST

RECORDS ARE INCLUDED IN REPORT IF REM FLOAT IS LESS OR EQUAL 10 DAYS.

SORT #5 CONTRACT, AREA, ASCENDING REMAINING FLOAT

AREA	ACCT	RES	I-T	I-J	DUR	PCT COMP	ACTIVITY DESCRIPTION	SCHED START	EST START	ACT START	EST COMPL	FLOAT TOT REM	CRD COE
789			01234	56789	2	2		D M Y	2 3	456789	012345		7 8
					0123	4567			890123	D M Y	4		890
									D M Y		D M Y		
030	000000	2	3506	3505	0005	095	EXC TUNNEL	170571		200571	160771	13 -26	C53
030	000000	1	8003	3004	0020	090	EXC PUMP HSE	170571		240571	190771	0 -25	C53
030	000000	2	3505	3575	0005	090	EXC TUNNEL	240571		270571	160771	13 -21	C53
030	100000	0	3004	3005	0010	050	CONC FTG 50	150671		210671	220771	0 -18	C53
030	000000	2	3575	3576	0005	088	EXC TUNNFL	010671		070671	160771	13 -16	C53
030	100000	0	3576	3577	0005	070	EXC TUNNEL	080671		140671	190771	13 -12	C53
030	100000	0	3006	3008	0015	000	CONC WALLS 50	290671			050871	0 -12	C53
030	100000	2	3005	3007	0010	000	CONC FTG 50	290671			290771	5 -7	C53
030	000000	1	3577	3578	0005	060	EXC TUNNEL	150671		210671	190771	10 -1	C53
030	100000	0	7012	3558	0002	000	BAC FILL INSIDE 50	110671			150871	0 -1	C53
030	100000	0	3557	3507	0010	085	TUNNEL SLAB	150671		210671	190771	13 -1	C53
030	100000	0	3506	3534	0010	095	TUNNEL SLAB	240571		210571	160771	28 0	C53
030	100000	0	3545	3546	0010	098	TUNNEL SLAB	080671		080671	160771	23 0	C53
030	100000	0	3008	3010	0015	000	CONC WALLS 50	210771		140671	160771	10 0	C53
030	100000	0	3012	3014	0003	000	CONC GRD SLAB	130671				0 0	C53
030	100000	0	3558	3559	0015	000	TUNNEL WALLS	290671				13 -1	C53
030	100000	0	3578	3579	0010	040	TUNNEL SLAB	220671		280671	050871	10 5	C53
030	500000	7	3605	3606	0060	000	TAILINGS LINE	170571	010672		230871 230872	269 6	C53
030	000000	1	3011	3013	0010	000	BAC FILL OUTSIDE	110671				8 8	C53
030	100000	0	3546	3547	0015	055	TUNNEL WALLS	220671		260671	260771	10 9	C53

Fig. D–9. Engineering – Construction Critical Activities List.

Source Holmes & Narver Inc., Orange, California, USA.

HOLMES & NARVER, INC.
ENGINEERS—CONSTRUCTORS

ENGINEERING PROGRESS REPORT SUMMARY

CLIENT_____ MANUFACTURING CO.   REPORT NUMBER_____ 3 _____

CONTRACT_____ 1440.00 _____   PERIOD ENDING ____ 01-28-77 ____

DESCRIPTION	SCHEDULED PERCENT COMPLETE	ACTUAL PERCENT COMPLETE	WEIGHTED FACTOR	WEIGHTED PERCENT COMPLETE
TASK I	100	100	25.74	31.76
TASK IA				
ARCHITECTURAL	34	50	4.44	2.21
MECHANICAL ENGR	34	10	6.17	.62
STRUCTURAL ENGR	34	15	2.26	.34
ELECTRICAL ENGR	34	30	7.41	2.22
TASK II				
ARCHITECTURAL	100	97	8.57	8.31
MECHANICAL ENGR	61	67	14.91	9.99
CIVIL ENGR	100	81	3.51	2.84
STRUCTURAL ENGR	100	96	10.10	9.70
ELECTRICAL ENGR	31	76	16.23	12.35
TOTAL PROJECT PROGRESS TO DATE	69			80
TOTAL PROJECT PROGRESS LAST PERIOD	56			65
TOTAL PROJECT PROGRESS THIS PERIOD	13			15

Fig. D–10   Engineering Detailed Progress Report.

Source   Holmes & Narver Inc., Orange, California, USA.

HOLMES & NARVER, INC.
ENGINEERS • CONSTRUCTORS

CONSTRUCTION PROGRESS REPORT SUMMARY

CLIENT_____MANUFACTURING CO._____   REPORT NO._____33_____
CONTRACT_____1440_____   PERIOD ENDING____11-20-77_____

ACCOUNT	ACCOUNT DESCRIPTION	SCHEDULED PERCENT COMPLETE	ACTUAL PERCENT COMPLETE	WEIGHTED FACTOR	WEIGHTED PERCENT COMPLETE
00	SITE PREPARATION	99	98	2.6	2.5
01	CONCRETE	100	100	7.4	7.4
02	STRUCTURAL	99	96	6.8	6.5
03	BUILDINGS	100	99	10.1	10.0
04	MACHINERY & EQUIPMENT	91	93	17.9	16.7
05	PIPING	98	99	20.1	20.0
06	ELECTRICAL	95	94	30.9	29.0
07	INSTRUMENTATION (WITH PIPING & ELECTRICAL)	-	-	-	-
08	PAINTING & INSULATION	66	78	2.6	2.0
09	INDIRECT FIELD OPERATIONS	70	71	1.6	1.1
	TOTAL PROJECT PROGRESS TO DATE	94			95
	TOTAL PROJECT PROGRESS LAST PERIOD	89			89
	TOTAL PROJECT PROGRESS THIS PERIOD	5			6

Fig. D–11   Construction Detailed Progress Report.
Source   Holmes & Narver Inc., Orange, California, USA.

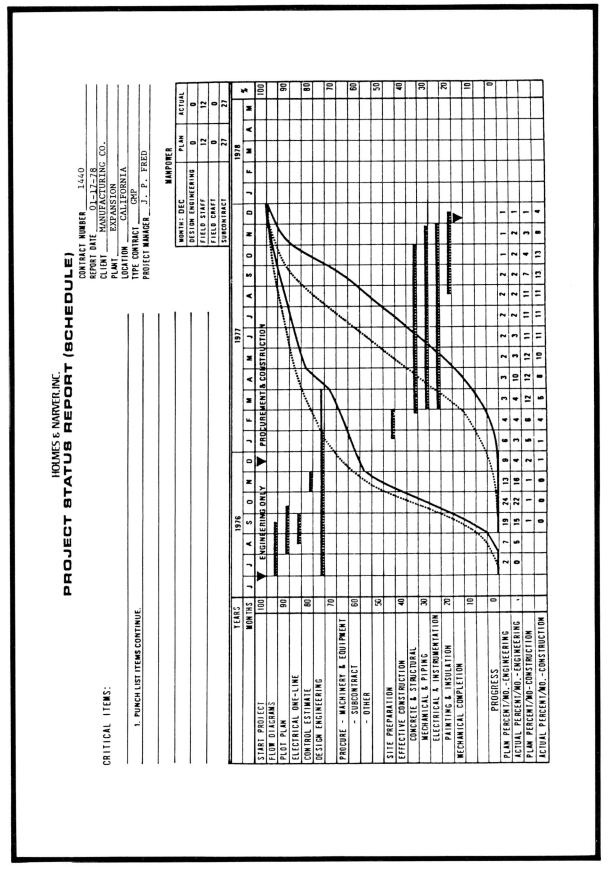

Fig. D–12 Project Status Report Schedule.

Source Holmes & Narver Inc., Orange, California, USA.

APPENDIX E

Cost Control tools

The tools illustrated and the definitions quoted below are reproduced from the Holmes & Narver Inc. Control System.

Fig. E–1. Project Code of Accounts. The preliminary budget estimate is disseminated into the Project Code of Accounts. This code is developed by the Project Manager and the Cost Engineer in consultation with the client.

Fig. E–2. Definitive Cost Estimate. With the essential completion of the design, the start of construction work and the ordering of major equipment, the initial budget estimate is replaced by the Definitive Cost Estimate which is the final budget control estimate of the project unless changes of scope are authorised by the client.

Fig. E–3. Engineering Labour Reports. Home office labour costs are recorded for all types of engineering work, procurement, etc., and are incorporated in a monthly labour cost and to-date figures report.

Fig. E–4. Field Labour Reports. Field labour costs disseminated for all construction accounts are incorporated in a monthly cost report similar to the home office report.

Fig. E–5. The Project Cost Summary shown in Fig. E–5 is issued monthly for the duration of the project. The report will contain overall composite project costs by prime cost account. The costs listed in the project cost summary are compiled from data accumulated in three subsidiary reports: namely, Engineering Labour Reports, Field Labour Reports and a Commitment Register of Purchase Order and Non-Purchase Order Expenditures. The Project Cost Summary Report states the job cost elements in columns headed 'Budget', 'Expended and Open Commitments', 'Indicated Total Cost', and 'Over or Under Budget'. Before the Definitive Estimate has been prepared, the budget estimate costs are set up by cost code account for use in the 'Budget' column of the Project Cost Summary. When the client approves a change order affecting the total price of the contract, the 'Budget' column of the Project Cost Summary will be adjusted accordingly.

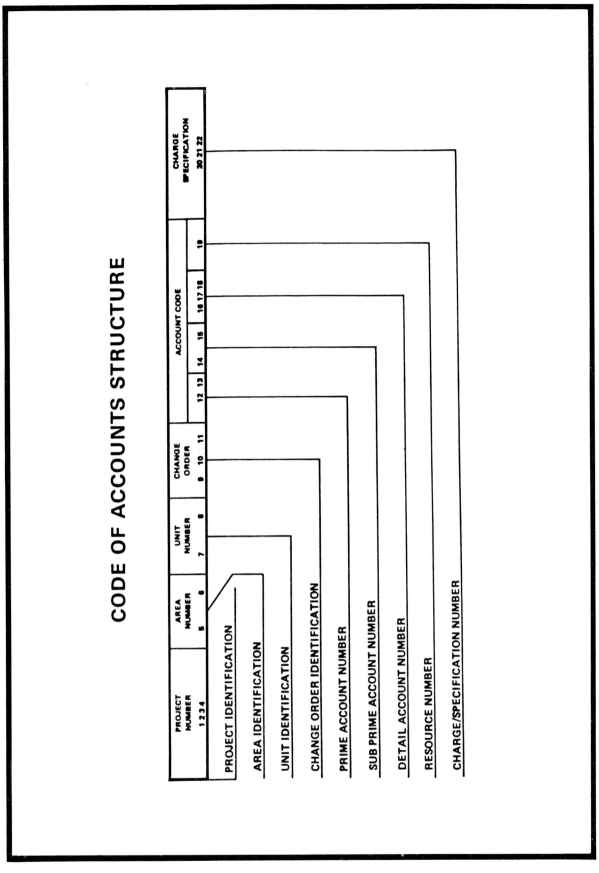

Fig. E–1 Project Code of Accounts
Source Holmes & Narver Inc., Orange, California, USA.

CUSTOMER _____

LOCATION _____

PROJECT _____

COST ESTIMATE

WO NO _____

MADE BY _____

APPROVED _____

A C NO	ITEM & DESCRIPTION	MANHOURS	ESTIMATED COST			
			LABOR	SUB CONTRACTS	MATERIALS	TOTAL
000	Site Preparation		400,000	500,000	200,000	1,100,000
100	Concrete		800,000	–	300,000	1,100,000
200	Structural Steel		100,000	1,200,000	200,000	1,500,000
300	Buildings		100,000	500,000	100,000	700,000
400	Machinery & Equipment		200,000	–	4,000,000	4,200,000
500	Piping		500,000		1,100,000	1,600,000
600	Electrical		300,000	700,000	400,000	1,400,000
700	Instrumentation		100,000	–	200,000	300,000
800	Painting & Insulation		–	100,000	–	100,000
	DIRECT FIELD COSTS	235,000	2,500,000	3,000,000	6,500,000	12,000,000
	Temporary Construction Facilities		400,000	–	100,000	500,000
	Construction Support Services		700,000	–	900,000	1,600,000
	Construction Equipment & Tools		–	300,000	100,000	400,000
	Special Items		–	–	–	–
	INDIRECT FIELD COSTS	35,000	1,100,000	300,000	1,100,000	2,500,000
	TOTAL FIELD COSTS					14,500,000
	Project Management		200,000			
	Design Engineering		700,000			
	Planning & Controls		100,000			
	Procurement		100,000			
	Office Expense					
	Office Payroll Burdens					
	Indirect Office Costs					
	TOTAL OFFICE COSTS	120,000	1,100,000		1,400,000	2,500,000
	TOTAL FIELD & OFFICE COSTS					17,000,000
	Fee					800,000
	Contingency					800,000
	Escalation					1,400,000
	TOTAL					20,000,000

Fig. E–2 Definitive Cost Estimate.
Source Holmes & Narver Inc., Orange, California, USA.

94

HOLMES & NARVER INC.

HOME OFFICE SUMMARY MANHOURS AND COST FORECAST

CLIENT: MANUFACTURING CO.
LOCATION: CALIFORNIA
PROJECT: EXPANSION

CONTRACT: 1400
DATE: 2-21-77
PERIOD ENDING: 2-11-77

DESCRIPTION	BUDGET MAN HOURS	BUDGET WAGE RATE	BUDGET DOLLARS	EXP MANHOURS THIS PERIOD MAN HOURS	EXP MANHOURS THIS PERIOD % BDGT	EXP MANHOURS TO DATE MAN HOURS	EXP MANHOURS TO DATE % BDGT	EXP WAGE THIS PERIOD	EXP WAGE TO DATE	EXP DOLLARS THIS PERIOD DOLLARS	EXP DOLLARS THIS PERIOD % BDGT	EXP DOLLARS TO DATE DOLLARS	EXP DOLLARS TO DATE % BDGT	FCST TO COMPL	FCST MANHOURS FORECAST TOTAL	FCST MANHOURS % OF BDGT	FCST MANHOURS OVER/UNDER HR BUDGET	FCST MANHOURS % BD	FCST WAGE TO COMPL	FCST DOLLARS TO COMPL	FCST DOLLARS FORECAST TOTAL	FCST DOLLARS % BDGT	FCST DOLLARS OVER/UNDER BUDGET
.77 882 ARCHITECTURE	2,782	10.16	28,273	50	2	1,880	68	8.16	9.63	408	1	18,104	64	530	2,410	87	(372)	78	9.65	5,115	23,219	82	(5,054)
883 MECHANICAL ENGR.	2,763	11.08	30,619	134	5	2,757	100	10.62	10.43	1,424	5	28,756	94	341	3,098	112	335	89	10.45	3,563	32,319	106	1,700
884 CIVIL ENGR	1,141	9.95	11,358	123	11	651	57	11.67	11.66	1,435	13	7,591	67	89	740	65	(401)	88	11.70	1,041	8,632	76	(2,726)
885 SPECS & COORD CK	503	8.60	4,325	2	0	464	92	9.50	9.73	19	0	4,515	104	76	540	107	37	86	9.75	741	5,256	122	931
886 ELECTRICAL ENGR	2,318	11.45	26,537	29	1	1,497	65	3.75	9.84	109	0	14,730	56	285	1,782	77	(536)	84	9.90	2,822	17,552	66	(8,985)
888 STRUCTURAL ENGR	1,393	12.62	20,100	30	2	1,472	92	14.43	12.51	433	2	18,415	92	182	1,654	104	61	89	12.55	2,284	20,699	103	599
TOTAL ENGINEERING	11,100	10.92	121,212	368	3	8,721	78	10.40	10.56	3,828	3	92,111	76	1,503	10,224	92	(876)	85	10.36	15,566	107,677	89	(13,535)
1W271 PROJECT MANAGEMENT	2,024	12.50	25,300	0	0	106	5	0	16.83	0	0	1,784	7	22	128	6	(1,896)	83	16.83	371	2,155	9	(23,145)
772/D8 PLANNING & CONTROLS	665	12.53	8,330	24	4	404	61	12.71	12.99	305	4	5,248	63	135	539	81	(126)	75	13.00	1,755	7,003	84	(1,327)
873A BC ADMIN/EXPEDITING	0	0	0	0	0	7	0	0	10.57	0	0	74	0	3	10	0	10	83	10.60	32	106	0	106
473 PURCHASING	80	7.56	605	0	0	2	3	0	4.50	0	0	9	1	3	5	6	(75)	83	4.50	14	23	4	(382)
881 PROJECT MANAGEMENT	1,625	14.53	23,615	100	6	1,039	64	15.28	15.28	1,528	6	15,876	67	228	1,267	78	(358)	82	15.30	3,488	19,364	82	(4,251)
889 PROJECT ENGINEERING	0	0	0	0	0	78	0	0	13.32	0	0	1,039	0	17	95	0	95	82	13.35	227	1,266	0	1,266
079B STENO/CLERICAL	518	5.50	2,849	35	7	477	92	5.23	5.22	183	6	2,490	87	98	575	111	57	83	5.25	515	3,005	105	156
TOTAL MANAGEMENT & SUPPORT	4,912	12.36	60,699	159	3	2,113	43	12.68	12.55	2,016	3	26,520	44	506	2,619	53	(2,293)	80	12.65	6,402	32,922	54	(27,777)
POINT NO. 77 TOTALS	16,012	11.36	181,911	527	3	10,834	68	11.09	10.95	5,844	3	118,631	65	2,009	12,843	80	(3,169)	83	10.93	21,968	140,599	77	(41,312)
.77 ENGINEERING	1,890	10.56	19,958	52	3	3,007	159	7.85	10.69	408	2	32,145	161	0	3,007	159	1,117	100	10.69	0	32,145	161	12,187
MANAGEMENT & SUPPORT	556	10.54	5,861	105	19	708	127	11.52	9.57	1,210	21	6,776	116	0	708	127	152	100	9.57	0	6,776	116	913
POINT NO. 77 TOTALS	2,446	10.56	25,819	157	6	3,715	152	10.31	10.48	1,618	6	38,921	151	0	3,715	152	1,269	100	10.48	0	38,921	151	13,102

Fig. E–3   Engineering Labour Reports.

Source   Holmes & Narver Inc., Orange, California, USA.

HOLMES & NARVER. INC.

FIELD LABOR SUMMARY
MANHOURS AND COST

CLIENT Salty Copper Company
LOCATION Arizona
PROJECT SX-EW Plant

CONTRACT 1976
DATE 10/3/75
PERIOD ENDING 9/30/75

A/C	DESCRIPTION	PHYSICAL PERCENT COMPLETE	MANHOURS SPENT TO DATE	MANHOURS BUDGETED	BUDGETED MANHOURS PER 1% COMPLETION	TO DATE ACTUAL MANHOURS PER 1% COMPLETE	FORECAST TO COMPLETE MANHOURS PER 1% COMPLETE	FORECAST TOTAL MANHOURS	MANHOURS OVER/UNDER BUDGET	FORECAST PERCENT PRODUCTIVITY (1)	CUMULATIVE PERCENT PRODUCTIVITY (2)	AVERAGE WAGE RATE BUDGETED	FORECAST TO COMPLETE AVERAGE WAGE RATE	DOLLARS TO DATE	DOLLARS BUDGETED	FORECAST DOLLARS	DOLLARS OVER/UNDER BUDGET	REMARKS
030	Earthwork	81	5,659	6,750	67	70	65	6,894	144	98	96	11.20	11.27	62,387	75,600	76,300	700	
051	Formwork	68	8,012	9,010	90	118	100	11,212	2,202	80	76	12.24	12.50	100,816	110,300	140,800	30,500	
054	Reinforcing Steel	60	1,719	1,850	19	29	20	2,519	669	73	66	13.54	14.97	27,127	25,100	39,100	14,000	
055	Concrete	55	2,238	5,500	55	41	50	4,488	(1,012)	123	134	11.00	11.58	28,946	60,500	55,000	(5,500)	
110	Rough Carpentry	-0-	-0-	306	3	-	3	306	-	100	100	12.42	12.42	-	3,800	3,800	-	
160	Misc. Metal Work	-0-	5	1,924	19	-	19	1,924	-	100	100	13.41	13.41	75	25,800	25,800	-	
270	Fire Protection System	2	8	378	4	4	4	378	-	100	100	16.77	16.73	110	6,300	6,330	25	
290	Prefab Buildings	100	52	50	1	1	-	52	2	96	100	12.48	-	649	624	649	25	
300	Electrical	90	56	60	1	1	-	62	2	97	100	11.50	12.50	640	690	715	25	
400	Water Utilities	100	51	40	-	-	-	51	3	94	100	15.90	15.94	810	761	810	47	
420	Sewerage Utilities	65	398	443	4	6	4	538	95	82	67	15.90	15.94	5,418	7,000	7,650	650	
500	Process Equipment	40	931	2,393	24	23	22	2,251	(142)	106	104	16.73	14.95	12,566	40,000	32,300	(7,700)	
600	Piping	16	1,102	7,582	76	69	65	6,562	(1,020)	116	110	16.74	17.35	18,715	126,400	113,466	(12,934)	
700	Instrumentation	-0-	-0-	1,951	20	-	20	1,951	-	100	100	16.76	16.76	-	32,700	32,700	-	
	TOTAL D.F. LABOR	50	20,231	38,245	383	405	379	39,188	943	98	95	13.48	14.62	254,279	515,577	535,390	19,813	

(1) FORECAST % PRODUCTIVITY = BUDGET MH x 100 / FORECAST TOTAL MH

(2) CUMULATIVE % PRODUCTIVITY = BUDGETED MANHOURS PER 1% COMPLETE x 100 / TO DATE ACTUAL MANHOURS PER 1% COMPLETE

PREPARED BY D.H.J.
SHEET NO. 4 of 7

Fig. E–4 Field Labour Reports.
Source Holmes & Narver Inc., Orange, California, USA.

RUN DATE 11/06/77
PERIOD ENDING 10/31/77

CLIENT- AMCO
PROJECT NO. 604   AREA 00   UNIT

HOLMES AND NARVER
PROJECT COST SUMMARY REPORT
PRIME ACCOUNT/BY RESOURCE
TOTAL PROJECT

ACCOUNT CODE	ACCOUNT DESCRIPTION	BUDGET	EXPENDED THIS PERIOD	EXPENDED TO DATE	EXPENDED AND OPEN COMMITMENTS	INDICATED TOTAL COST	OVER/UNDER BUDGET	PCT
00	SITE PREPARATION AND CIVIL	3,599,433	12,254	3,338,011	3,338,011	3,600,124	691	
01	CONCRETE	20,273,454	31,010	17,217,914	17,217,914	19,762,315	-511,139	-3
02	STRUCTURAL	6,707,371	24,675	5,210,710	5,210,710	6,707,371		1
03	BUILDINGS	8,038,885	39,669	6,510,798	6,510,798	8,121,418	82,533	-1
04	MACHINERY AND EQUIPMENT	21,919,578	84,719	16,904,342	16,904,342	21,899,302	-20,276	
05	PIPING	26,233,219	121,914	19,417,553	19,417,553	25,972,408	-260,811	-1
06	ELECTRICAL	10,143,042	104,196	6,331,419	6,331,419	10,143,042		
07	INSTRUMENTATION	2,192,465	36,462	1,426,566	1,426,566	2,197,013	5,948	
08	PAINTING	249,779	9,640	129,578	129,578	247,217	-2,512	-1
09	INSULATION	166,152	7,294	93,133	93,133	166,152		
	DIRECT FIELD LABOR	99,523,328	475,833	76,580,024	76,580,024	98,817,162	-706,166	-1
00	SITE PREPARATION AND CIVIL	4,970,567	19,168	4,803,481	4,803,481	5,031,351	60,784	1
01	CONCRETE	27,996,546	48,506	24,776,999	24,776,999	26,540,199	-1,456,347	-5
02	STRUCTURAL	9,262,629	39,080	8,044,390	8,044,390	9,262,629		
03	BUILDINGS	11,101,115	62,048	9,183,904	9,183,904	13,097,501	1,996,386	18
04	MACHINERY AND EQUIPMENT	30,270,422	129,991	24,325,762	28,458,859	29,308,131	-962,291	-3
05	PIPING	36,226,781	190,670	24,676,650	28,676,650	34,436,709	-1,790,072	-5
06	ELECTRICAL	14,006,958	169,205	9,125,394	9,125,394	14,006,958		
07	INSTRUMENTATION	3,027,535	57,036	2,070,840	2,070,840	3,256,503	228,968	8
08	PAINTING	344,771	15,078	185,507	185,507	330,186	-14,585	-4
09	INSULATION	229,448	11,411	136,315	136,315	229,448		
	DIRECT FIELD MATERIAL	137,436,772	742,189	111,329,242	115,462,339	135,499,615	-1,337,157	-1

Fig. E–5   Project Cost Summary.

Source   Holmes & Narver Inc., Orange, California, USA.

97

APPENDIX F

Value Engineering

During World War II, at the General Electric Company, Lawrence Miles developed a system of techniques to reduce costs and improve products which he called Value Analysis. Today, the two terms Value Engineering (V.E.) and Value Analysis (V.A.) are used synonymously. Terms such as Value Control and Value Management are also used.

For the construction industry in the USA, the concept was introduced by Alphonse Dell'Isola into the Navy Facilities Engineering Command in 1963. The US Army Corps of Engineers followed with a program in 1965.

Value Engineering is a practical management tool that is able to analyse and evaluate changes in order to aid in developing maximum benefit from available resources. Alphonse Dell'Isola defined it as 'a creative organised approach whose objective is to optimise cost and performance of a facility or system.'

Fig. F–1. Illustrates cost reduction potential vs. Cost to Change by using V.E.

Fig. F–2. Illustrates a Total value Management Program.

Fig. F–3. Illustrates the phases of the Value Analysis/Engineering Job Plan.

Fig. F–4. Illustrates the Methodology of the V. E. Job Plan.

Fig. F–5. Illustrates the Value Engineering Procedures.

[1] Alphonse Dell'Isola, *Value Engineering in the Construction Industry*, New York, 1982

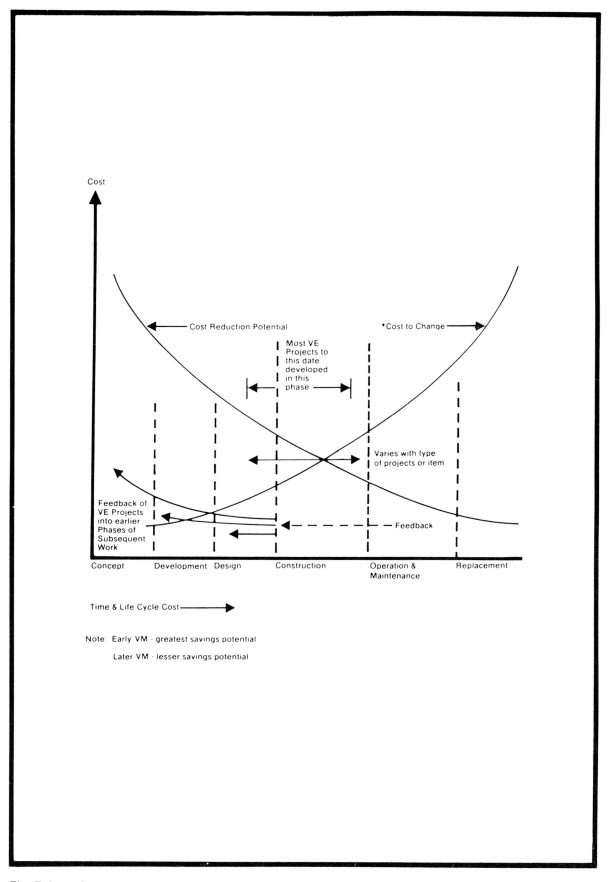

Fig. F-1   Cost Reduction Potential vs. Cost to Change.

Source   Alphonse Dell'Isola, *Value Engineering in the Construction Industry*, New York, 1982.

A TOTAL VALUE MANAGEMENT PROGRAM

ITEM	ACTION	WHEN	PURPOSE/RESULT
- Value Control	Function analysis of Context Planning and user's needs.		Identify baseline performance and excessive costs. Plan for Profit.
- Value Engineering.	Value Engineering study of preliminary design. Correct direction.	At 30% design stage	Avoid cost & time growth in skills, materials, methods & Equipment.
- Value Analysis	Value Engineering study of completed design	Pre-contract & specification preparation.	Update alternate materials and methods that are acceptable.
- Value Engineering change proposals	Encourage suppliars and contractors to use VECP Clause	Pre-construction conferences.	Accumulate new ideas from specialty areas 5-10% saving.
- Value Engineering of operation & Maintenance	Value Engineering using O & M Specialists	Prior to occupancy	Maximize LCC O & M Savings

Fig. F–2   Total Value Management Programme.
Source   A Lecture by William L. Kelly, Riyadh, 1983.

100

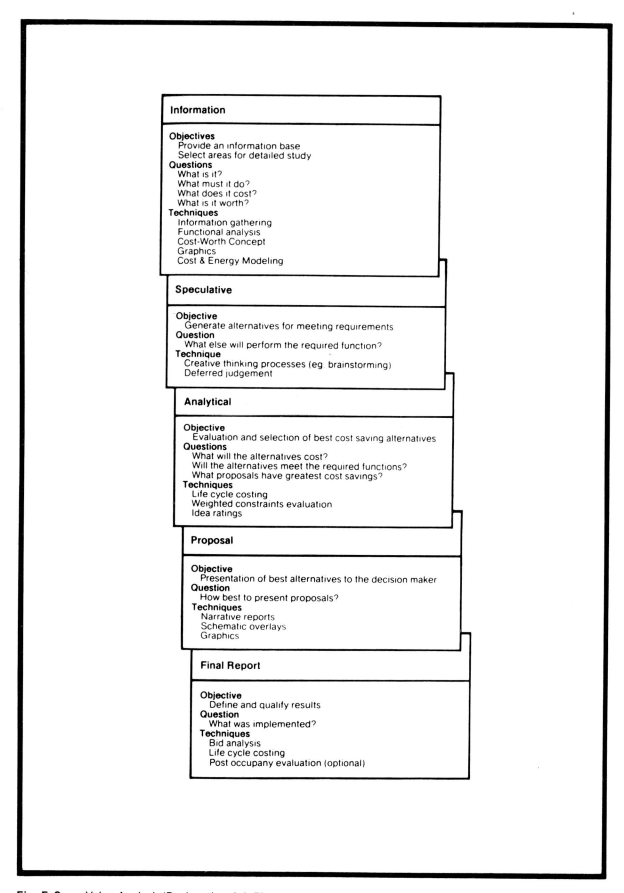

Information

Objectives
　Provide an information base
　Select areas for detailed study
Questions
　What is it?
　What must it do?
　What does it cost?
　What is it worth?
Techniques
　Information gathering
　Functional analysis
　Cost-Worth Concept
　Graphics
　Cost & Energy Modeling

Speculative

Objective
　Generate alternatives for meeting requirements
Question
　What else will perform the required function?
Technique
　Creative thinking processes (eg. brainstorming)
　Deferred judgement

Analytical

Objective
　Evaluation and selection of best cost saving alternatives
Questions
　What will the alternatives cost?
　Will the alternatives meet the required functions?
　What proposals have greatest cost savings?
Techniques
　Life cycle costing
　Weighted constraints evaluation
　Idea ratings

Proposal

Objective
　Presentation of best alternatives to the decision maker
Question
　How best to present proposals?
Techniques
　Narrative reports
　Schematic overlays
　Graphics

Final Report

Objective
　Define and qualify results
Question
　What was implemented?
Techniques
　Bid analysis
　Life cycle costing
　Post occupancy evaluation (optional)

Fig. F–3　　Value Analysis/Engineering Job Plan.
Source　　　　Alphonse Dell'Isola, *Value Engineering in the Construction Industry*, New York, 1982.

Objectives	Job Plan	Key Techniques	Supporting Techniques	VE Questions
	Information phase	Get all facts Determine cost and/or quantities	Obtain all information Work on specifics	What it it? What does it cost? What amounts are used?
Define functions		Define function Put dollar value on specifications and requirements Determine worth Cost and energy models	Divide problem into functional areas	What is its function? What is the value of the function? What are the isolated areas for study?
Create ideas	Speculative phase	Blast and create	Create Innovate Defer judgement	What else will perform the function?

Evaluate basic function	Analytical phase	Evaluate:
		  Basic function
		  By comparison
			Evaluate functional areas	What ideas will perform the function?

Evaluate new ideas		Quantify and put dollar value on ideas and refine	Analyze cost, use good judgement	

Consult		Investigation
		  Suppliers
		  Companies
		  Consultants
			Investigate advanced techniques	What else will do the job?

Compare		Use Standards
		Compare:
		  Methods
		  Products
		  Materials
			Develop new ideas	

Develop alternates		Determine costs	Use teamwork	What will alternates cost/or use?
List best ideas	Proposal phase	Extract data	Use good human relations	What alternates are recommended for implementation?
Summarize		Motivate positive action	Finalize solutions	
Document			Document and present solutions for action	
Implement ideas Validate results	Final report phase	Check bid prices Post occupancy evaluation (optional)	Life cycle costing On-site inspection	What ideas were implemented? What savings were realized?

Fig. F–4   Methodology of the V.E. Job Plan.

Source   Alphonse Dell'Isola, *Value Engineering in the Construction Industry*, New York, 1982.

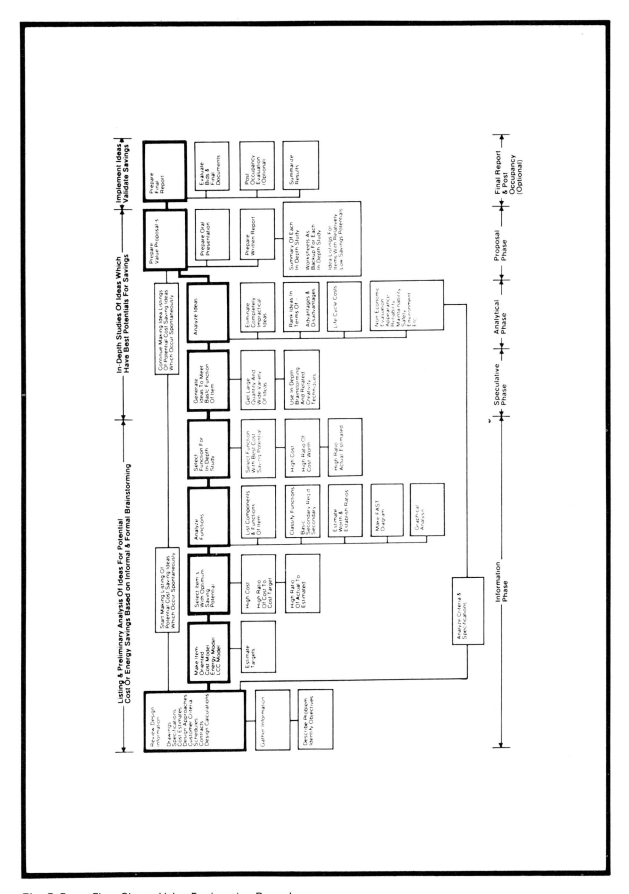

Fig. 5–5   Flow Chart – Value Engineering Procedures.

Source   Alphonse Dell'Isola, *Value Engineering in the Construction Industry*, New York, 1982.

103

APPENDIX G

Training

An example of a training course conducted by the Urwick Management Centre, Slough, England. As indicated in the description of this course, the programme includes a high proportion of practical work. In addition, course members are encouraged to discuss their own work experiences and receive personal help from the Course Director and other tutorial staff. Extensive use is made of exercises, case studies and films. Course members are given a comprehensive set of tutorial notes covering all aspects of the course programme. These are retained by course members after the course and form a valuable work of reference for use on return to their work.

Managing Large-Scale Projects

Introduction

This new 10-week course has been designed to give a thorough grounding in project management. The emphasis is on project implementation – effective management of the many complex tasks which must be carried out between project authorisation and completion.

The programme will be of help to all those involved in large scale projects. It is designed and conducted in the U.K. by the Urwick Management Centre – the training division of international management consultants Urwick, Orr and Partners Ltd. Participants will benefit from the knowledge and practical experience of tutors who are specialists in applying what they teach in many countries throughout the world.

This course is especially suitable for managers from developing countries who may be eligible for financial awards. Further details are given on the back of this leaflet.

Course Objectives

- To give comprehensive instruction on the principles and techniques of project management as applied to major projects such as: industrial, infrastructure, public works, transportation, power generation, communications, marine projects, etc.
- To provide practical advice and assistance to managers on the *implementation* of projects, particularly in respect of project planning, organisation and control.
- To help participants to improve their managerial effectiveness in areas such as project investigation, reporting, achieving improvements, resolving problems with contractors, agencies and others involved in their projects.

Intended For

- Senior and middle managers involved in large-scale projects, who wish to improve their proficiency in project implementation.
- Technical specialists, such as engineers or surveyors, who are assuming a broader *management* role in projects.
- Senior officers in government or administrative agencies responsible for monitoring and controlling major capital projects.
- Managers working for consultants or contractors providing services to projects.
- Teachers of management who wish to become competent in this field.

Course Structure

The course is divided into four modules:

1 Principles of project management	-	4 weeks
2 Site investigation and reporting	-	3 weeks
3 Company visits	-	1 week
4 Successful implementation	-	2 weeks

The main part of the programme will be conducted at the Urwick Management Centre, Slough, England. Practical work and visits to companies engaged in large-scale projects occupy approximately 3 weeks during the middle of the course and will take place at various locations in the U.K.

Course Content

1 Principles of Project Management

This module covers the main stages of a project and examines proven management techniques relevant to each stage. Subjects include:

- Project framework
- Appraisal and approval
- Programming and scheduling
- Financial planning
- Reporting and information systems
- Managing design
- Procurement
- Forms of contract
- Contractual relationships
- Construction phase
  - planning and control • use of computers
  - cost control • payments, variations
  - documentation • industrial relations
  - safety
- Commissioning
  - investigational techniques • methods of reporting • effective teamwork

2 Site Investigation and Reporting

Participants carry out in-depth studies of a number of important aspects of project management at selected sites in the U.K. On return, they prepare and make presentations of their findings to tutors and other course members.

3 Company Visits

Visits will be arranged to companies engaged in major projects to enable course members to:

- broaden their experience
- discuss practical aspects of project implementation with managers and staff
- further improve their diagnostic skills
- pursue topics related to their own work and experience

4 Successful Implementation

The emphasis in this module is on 'how to do it in practice'. Participants will be helped to make use of knowledge gained in the course, tackle project problems, improve project performance, overcome resistance to change, deal with contractors, government agencies, etc., plan and organise their personal time.

At the end of the course each participant will have produced a personal action plan for implementation on return to work.

Urwick Management Centre
Baylis House
Stoke Poges Lane
Slough, Berks
SL1 3PF
U.K.

urwick

Fig. G–1 A Project Management Course.
Source Urwick Management Centre, UK.

INDEX